貿易英文
Trade English

李再福 著

五南圖書出版公司 印行

作者簡歷

李再福

1933, 2　出生於台灣省台南市

1958, 6　國立台灣大學經濟系畢業

1958, 8～1960, 2　成功大學會統系助教

1960, 2　進中央信託局,歷經壽險處、信託處及外匯業務處之辦事員、副主任、主任、襄理及副理。

1987, 3　外匯業務處經理

1998, 7　屆齡退休

曾 兼 任：淡水工商管理專科學校企業管理科講師(1969, 9～1971, 7)
　　　　　明志工業專科學校工業管理科講師(1975, 8～1976, 7)
　　　　　實踐設計管理學院會統科講師(1976, 8～1992, 7)
　　　　　中國大陸災胞救濟總會職業訓練所教師(1983, 4～1986, 6)

現 　　任：金融人員研究訓練中心,外匯業務訓練班教師

著 　　作：貿易英文　經世文化事業股份有限公司(1982, 8初版,1992, 11修訂)

著 　　述：從出口結匯談信用狀　中信通訊118期(1978, 5)
　　　　　撰寫貿易書信的要訣　中信通訊127期(1980, 7)
　　　　　1983年信用狀統一慣例簡述　中信通訊142期(1984, 5)
　　　　　淺談新舊信用狀統一慣例(上)　中信通訊　行政革新專刊(1994, 1)
　　　　　淺談新舊信用狀統一慣例(下)　中信通訊194期(1994, 4)
　　　　　信用狀項下單據之審查與信用狀統一慣例　中信通訊203期(1996, 7)
　　　　　淺談SWIFT開發的信用狀　中信通訊210期(1998, 4)
　　　　　信用狀交易Q&A　中信通訊212期(1998, 10)

新版序

時代在轉變，對外貿易的作業也受環境的變遷，尤其是電腦連線及網際網路的普及而有所調整。本書於 1992 年 11 月再版之後不及五載即有修訂之必要，惟因某種特殊原因而停頓。幸蒙　五南圖書出版公司承接始得重獲翻修之良機。

經過半年多審慎的修正與補充，終於以嶄新的面貌和讀者再見面。茲舉重要修訂的內容如下：

一、版面翻新。

二、全書詞句辭語均逐頁推敲修正。

三、配合實務作業之變更，重繕部分甚多，其中要項如後：

　　㈠貿易英文的內容簡介；

　　㈡出進口廠商登記辦法；

　　㈢三 C 的定義；

　　㈣外匯管理有關之規定；

　　㈤1990 年修訂的國貿條規簡介；

　　㈥交貨時間的解釋；

　　㈦依 1993 年修訂之信用狀統一慣例，修正信用狀的定義；

　　㈧裝運程序；

　　㈨報關文件及

　　㈩通關程序等。

四、範例更新，尤其履行交易階段的有關表報，案例變更最多。

五、各章之後的練習問題均重新設計編排以利讀者複習增加學習效果。

六、為讓讀者因應自 1999 年 1 月 1 日起開始流通的新種貨幣「歐元」，補充「附錄六　認識歐元」供參考。

以上所舉各項請讀者連同上述再版之序及原序文一併參考，則必然能深入了解本書的架構和所欲闡明的主旨。

　　本書之能夠重新出版，除感謝五南圖書出版公司之鼎力支持之外，對協助提供資料的朋友和老同事，在此一併致萬分之謝意。至於本書雖然經過細心的推敲修正，惟恐尚有未逮之處，仍祈讀者不吝指教。

李再福　謹識

1998 年 12 月 21 日　於台北市

再版序

　　本書出版迄今已逾十年，承蒙企業界人士之肯定以及著者執教之大專院校學生的愛用，謹誌謝意。十年來世界經貿環境日新月異，國內的金融貿易政策與法令亦隨之大幅轉變而陸續修訂。

　　中央銀行於七十五年五月十四日修正「外匯管理條例」，將外匯重新定義為外國貨幣、票據及有價證券，並採行外匯申報制。七十六年七月十五日起外匯管理措施逐步放寬。諸如提高匯出入匯款限額、外匯存款、遠期外匯及匯率之議價等。而我國的外匯市場由原來的「外匯交易中心」改設「台北外匯市場發展基金會」擔任外匯買賣的仲介。貨品的輸出入管理方面，擴大免簽證項目範圍，並且申請書不再列入有關結匯事項。通關手續則簡化而更新報單。類此措施無不影響廠商對外貿易的作業。

　　在國際間，信用狀依然是主要的支付貨款的工具。但開狀方式以環球銀行財務電信協會(SWIFT)的電訊系統開發者愈多。同時進出口廠商之間的溝通方式逐漸加速。雖然使用傳真電報(FAX)已蔚然成風尚，相對地突出書信的重要性。蓋傳真仍須事先繕妥書信方能拍發。傳遞的訊息依然保持傳統的書信架構。

　　鑒於上述重大變革，著者花了近半年時間重校內容，更新資料。配以嶄新的單據，希望讀者藉再讀本書，不但因此獲得最新的貿易概念，而且有助於繕寫內容符合現代化的書信。此次修訂有關報關提貨部份經由捷凱通運公司業務經理許伯耀先生之協助，特此致謝。至於全書更動事項難免有疏漏之處，尚祈讀者不吝指正。

李再福　謹識

民國八十一年十一月於台北

Trade 序 nglish

　　本書針對大專肄業或即將畢業的學生，有意在未踏入社會之前，鞏固貿易英文的基礎，或貿易界在職人士，有志在短期內充實貿易英文，以應付業務上的需要，提供最扼要又實用的貿易英文的指引。

　　從事對外貿易或與之有關聯的事務，均離不開貿易英文。貿易英文涵蓋頗廣，基本上則以貿易書信為主。目前的貿易界，雖然使用電報交換(Telex)的機會愈來愈多，為了爭取時效，操作電報交換往往不容在紙上先作稿，而須隨時得將公司的意旨拍發對方。(培養)這種講究快速又通順的英文表達能力，需靠正確的貿易英文和熟練的業務經驗相互配合。

　　一般而言，具有大專畢業程度的人士，未必能寫出通順的貿易書信，而有實際貿易工作經驗者，又有視撰寫貿易書信為畏途。究其原因，不外前者缺乏實務經驗；後者則忽略了貿易英文的本質。貿易書信(即狹義的貿易英文)使用的英文有其獨特的語法和術語。這些語法和術語必須配以業務的流程適時地應用。因此學習貿易書信最有效的方法，依著者之見，首先須熟悉實務的環境，換言之，假設置身於其中，再將業務處理的構想，用業者所熟悉的語調，平易地敘述。

　　為此，著者將本身在外匯部門工作二十多年及公餘兼課十餘年的經驗，整理有關的講義，彙總相關的資料編成本書。其內容順著貿易的進行，述及實務的觀念和處理的程序，再舉例函講解構想和撰寫的原則。初學者依序而讀諒必有收穫，有工作經驗或目前從事於工作者，可以擇閱有關章節作為補充。

　　每章之後所附練習不限於復習內容，有不少挑戰性的問題，讀者亦可藉以涉獵有關書籍深入探討則必有心得。附錄一「貿易書信常用基本字彙」不但可供撰稿時的參考，亦可當貿易常識閱讀。附錄二「信用狀統一慣例」則為接觸信用狀交易者必備的參考資料，亦一併錄之。

　　本書付梓時間匆促，疏誤深恐難免，尚祈各界不吝指正為幸。本書之撰稿承蒙　學長張錦源教授之鼓勵與指導，以及幾位同事之幫忙，在此一併致謝。

李再福　謹識

民國七十一年八月

目錄

第十章　提貨與服務　205

附錄　225

第一章 Trade-English

緒　　論

一、貿易英文的意義

　　貿易英文源自英國商界用於溝通彼此以達到商業上某特定目的之英文。由於當年英國稱霸七大洋，發展對外貿易，以致貿易英文成為溝通國際間從事於貿易者的媒介。惟當時從英國本島經過遙遠的路途傳遞至海外各地的貿易書信，其功能僅止於達到通訊之目的。因此英語常以 Correspondence 一詞指貿易英文，而有關的書籍則取名為 " Commercial Correspondence "、" Business Correspondence "、" Foreign Trade Correspondence "、" Commercial English " 或 " English for Business " 等。二次大戰以後，世界的金融中心轉移至新大陸。國際間交易的主要貨幣亦由美金取代英鎊。幅員廣闊的美國不像英國需依賴對外貿易；為了配合廣大的國內消費市場，推行大量生產和大規模銷售。因而溝通的手段已不再局限於書信，更透過電訊網、視聽的廣告而發揮了說服的功能。因此美語以 Communication 一詞指貿易英文。是故，他們的書名常取名為 " Business Communication"、"Business English in Communication"、"Communication in English " 或 " Business Writing " 等。這種具有通訊和說服兩種功能之英文在我國通常稱為商用英文。惟國情之不同，一般國內的商業行為並不需要商用英文。只有從事對外貿易或與此有關的場合始使用，故宜稱之為貿易英文(Trade English or English for Foreign Trade)。

二、貿易英文的內容

　　貿易英文既然為從事對外貿易或與此有關的場合所使用，則其應用範圍包括了(1)貿易書信(Business or Commercial Letters)，即與國外或駐在國內的外國廠商或機構之間溝通的信函；(2)貿易電訊(Business Telecommunications)，即以電報(Telegram)、電報交換(Telex)或傳真(Fax)等方式溝通者；(3)各項貿易文件(Commercial Documents)，即對外貿易的作業上必須者，如商業發票(Commecial Invoice)、提單(Bill of Lading)、保險單(Insurance Policy)及合約(Contract or Agreement)等，以及(4)廣告文件(Publicity Materials)，即推銷產品時需用的規格清單(Specifications)、目錄(Catalogs)或小冊子(Brochures)等。

由於貿易書信乃對外貿易的實務上，最基本而且最經常處理的溝通工具，而且貿易電訊雖然採用不同的溝通方式，其文體的基本結構衍生自貿易書信，故謂學習貿易英文之基礎在於貿易書信亦不爲過。本書以下所稱貿易英文，除非特別聲明，係指貿易書信爲主。

三、貿易英文的特徵

1.有獨特的插語和慣用語

貿易英文反映從商者和氣生財的本質，常用獨特的插語和慣用語，以便柔和語氣。例如信函的開頭「八十七年十月十日大函敬悉」用平常的英文可以寫作 We received your letter of October 10，1998. 但是換成貿易英文則爲：

We **are glad** to receive your letter of October 10, 1998.

或 We **are pleased** to receive your letter dated October 10, 1998.

或 We **have the pleasure** to receive your letter dated October 10, 1998. 等。

句中粗體字部分乃具有潤滑作用的插語(Softeners)。在信的末段表達「敬請賜覆」的敘述句爲

We hope to receive your reply.

若利用插語，如粗體字部分則可以寫成

We **are waiting for** your reply.

或 We **shall appreciate** your reply.

或 We **look forward to** receiving your reply.

而使得整句腔圓舒適。

從事對外貿易需先建立業務往來。

We are writing to you with a view to **entering into business relations with** you.（謹特函以期建立業務往來關係）

句中粗體字部分乃表達「與……建立業務往來關係」的慣用語（Idiomatic Expressions）。同業間的往來建立於信用的基礎，為了解對方的信用，在致徵信機構委託徵信的函件裏，委託人必須具結如下：

Your information will be **treated as strictly confidential**.

（貴公司所提供資料當視為機密絕不公開）

句中粗體字部分即表示「視為機密絕不公開」的慣用語。

2.文中常用貿易術語(Trade Terminology)

使用貿易術語也是貿易英文的特徵之一，其目的在於簡化複雜的敘述並確定其意義。例如「有關本公司 TA－1 型立體卡式收錄音機，茲以到紐約港運費、保險費在內報最低價」，按一般的英文寫

We have quoted the lowest prices **including the cost of the goods, the insurance premium and the ocean freight to** New York for our Model TA－1 Stereo Radio Cassette Recorder.

雖然詞意通順但是浪費雙方寶貴的時間；蓋句中粗體字部分可以 CIF 代之而寫成

We have quoted the lowest prices CIF New York for our Model TA－1 Stereo Radio Cassette Recorder.

則不但簡潔而且在這種貿易條件下，買賣雙方的義務責任自有國際間通行的解釋可循，因而意義更為確切。

二次大戰以後，貿易英文免不了使用貿易術語之外，受了美語的影響，令人望而生畏的古板又拘束的插語或慣用語逐漸收斂。類似 We acknowledge receipt of your letter of January 20, 1998. 的句型已不多見，而代之以爽直的 Thank you for your letter dated January 20, 1998. 或 We received with thanks your letter dated January 20, 1998. 等句子。

四、學習貿易英文應有的基本英文

一般而言，學習語言必須聽、說、讀、寫並進。但是貿易英文屬於文語，故而比較偏重閱讀(Reading)和寫作(Writing)的能力，同時以實用的單字和活用的文法為基礎。由以下的測驗，讀者不難測出你目前的能力，而且可以領略貿易英文所需要的英文能力與範圍。

測驗一　你認識下列的字嗎？

1. account 　　2. exchange 　　3. protest
4. quality 　　5. value 　　6. insurance
7. deficit 　　8. recession 　　9. securities
10. speculation

測驗二　你能寫出下列的英文嗎？

1. 開發　　2. 設計　　3. 財產　　4. 浪費　　5. 重量
6. 決定　　7. 競爭　　8. 趨勢　　9. 升值　　10. 順差

測驗三　你認識下列的縮寫嗎？

1. GNP 　　2. B/L 　　3. IC
4. CY 　　5. GATT 　　6. BOFT
7. CFR 　　8. D/A 　　9. L/C
10. WTO

測驗一提示貿易英文常用的名詞，是屬於閱讀用的被動字彙(Passive Vocabulary)。測驗二則為寫作貿易英文常用的主動字彙(Active Vocabulary)。測驗三乃貿易英文常用的縮寫(以上的正確答案參閱第9頁)。

測驗四　下列一則報導，你能閱讀後立即掌握其內容嗎？

Britain Floats Pound

LONDON－－Britain floated the pound sterling on June 23

to head off a threatened devaluation.

The British move touched off a new monetary crisis and the European money markets were closed until early last week.

測驗五　請迅速閱讀下面的書信，你能立即回答有關的問題嗎？

Dear Sirs,

Thank you for your quotation of 28th June, 1998, and we note that the total cost of the 200 units is £50, 000. 00 CIF London.

We agree to this price and would ask you to accept this letter as our official order for the goods in question.

We will arrange the L / C upon receipt of your acknowledgement.

Yours faithfully,

問題一　誰寫這封信？賣方或買方？

問題二　運費和保險費包括在總金額嗎？

問題三　寫這封信的是中國廠商或英國廠商或美國廠商？

問題四　本案的貨款如何支付？

問題五　這筆交易成交了嗎？

測驗四、五包括了貿易英文經常接觸的內容。測驗四是有關通貨問題，不認識 float, pound sterling, devaluation 等關鍵字(key words)則無法把握主題，而 head off, touch off 等則乃輔佐文意的慣用語 (idiomatic expressions)。測驗五是一則貿易書信，可以測試你對貿易實務方面的認識程度 (以上正確答案參閱第 10 頁)。

測驗六　下列各句的意思表達你能辨別正確與否嗎？正確者用(√)號標出。

1. A（　）We are considering increasing prices.

   B（　）We are considering to increase prices.

   C（　）We are considering an increase in price.

2. A（　）We recommend using type carbon brushes.

   B（　）We recommend to use type carbon brushes.

   C（　）We recommend you to use type carbon brushes.

3. A（　）I explain you about this matter.

   B（　）I explain about this matter to you.

   C（　）I explain this matter to you.

4. A（　）The above price is not including your commission.

   B（　）The above price does not include your commission.

5. A（　）We will ship the goods in July, if there will be a steamer available.

   B（　）We will ship the goods in July, if there is a steamer available.

   C（　）We will ship the goods in July, if there was a steamer available.

測驗七　下列各句的意思你能利用指定的字彙寫出英文嗎？

1.回信太遲敬請原諒。

Excuse, in answering, for, me, my being, too late, your letter

2.快睡吧，明早還要早起呢。

Go, hard, or you will find it, at once, to bed, tomorrow morning, to get up early

3.你來此地後遇到過這樣寒冷的冬天嗎？

Have, a cold winter, ever, experienced, since you came

here, such, you, ?

4.你的鞋子在那裏修的？

Where, have, did, mended, shoes, you, your, ?

5.我們必須未雨綢繆。

We must, rainy, something, against, save, a, day

測驗六和七是寫作的基本原則。英文作文不能靠知識的文法，必須熟悉英文本身的語法(usage)。測驗六的 1. 2. 3.是動詞類型(Verb Patterns)的問題。雖然 I want you to do…是正確的表達，但是 I suggest you to do…卻是錯誤的，應該說 I suggest that you do…(以上的正確答案參閱第 10 頁)。

測驗題解答 •

測驗一　1.計算，帳戶　　2.交換，匯兌　　3.抗議，拒絕證書

　　　　4.品質　　　　　5.價值，價格　　6.保險

　　　　7.赤字，逆差　　8.不景氣，蕭條　9.有價證券

　　　　10.投機

測驗二　1)development　2)design　　　3)property

　　　　4)waste　　　　5)weight　　　6)decision

　　　　7)competition　8)tendency, trend

　　　　9)appreciation　10)surplus

測驗三　1)Gross National Product 國民總生產

　　　　2)Bill of lading 提單

　　　　3)Integrated Circuit 集積回路

　　　　4)Container Yard 貨櫃場

　　　　5)General Agreement on Tariffs and Trade 關稅及一般貿易協定

　　　　6)Board of Foreign Trade 國際貿易局

　　　　7)Cost & Freight 運費在內(貿易條件)

8) Documents against Acceptance 承兌交單

9) Letter of Credit 信用狀

10) World Trade Organization 世界貿易組織

測驗四　英鎊採浮動匯率

倫敦消息—英國於六月二十三日爲避免英鎊勢將貶值乃採取浮動匯率。英國的舉動觸發新的貨幣危機因而歐洲貨幣市場關閉至上週初。

測驗五　1.買方(Quotation 是賣方寄出的)

2.包括在內(CIF 報價)

3.英國的廠商(日期的寫法及報價的進口港是 London)

4.憑信用狀(最後一段提到 arrange the L/C)

5.尚未成交(本函尚須對方的 Acknowledgement)

測驗六　1)A，C　　2)A　　3)C　　4)B　　5)B

測驗七　1) Excuse me for my being too late in answering your letter.

2) Go to bed at once or you will find it hard to get up early tomorrow morning.

3) Have you ever experienced such a cold winter since you came here?

4) Where did you have your shoes mended?

5) We must save something against a rainy day.

練習問題

一、何謂貿易英文？

二、貿易英文的應用範圍爲何？

三、試述貿易英文的特徵。

四、貿易書信爲電文(Cable)和電報交換(Telex)及傳眞(FAX)的基礎。何故？

五、下列各句是貿易書信常見的語句，試指出其主旨(不必詳細翻譯)，並討論 1.誰寫的？ 2.爲何而寫？

1. The catalogue you requested on October 20 mailed today.

2. Here is the bulletin you asked us to send.

3. Thank you for your helpful suggestions about our sales conference.

4. We are pleased to send you the material you requested.

5. The material on page 16 of the enclosed brochure will answer the questions in your letter of December 15.

6. Just as soon as we received your letter, we wired our New York office to ship the goods.

7. The tires which complete your order LL−137 were shipped today.

8. Your letter of November 11 came as a surprise to me.

9. Many thanks for your letter of the 10th.

10. A copy of the annual report is being sent to you as requested in your letter of the 12th.

Trade English

第二章

貿易程序與貿易書信

貿易書信雖然繁多複雜，但是可以按其本文的內容加以分類研究之。蓋本文在貿易進行的各階段，有使用類似辭句之趨勢，內容類同的書信可冠以通稱，例如詢價函、報價函等。無適當名稱者，亦由該書信所屬貿易階段而得知其概略，是故要了解貿易書信的分類，宜先認識對外貿易的一般交易程序。本書將交易的程序分爲準備交易、進行交易、履行交易及售後服務等四個階段。在未談及各階段的程序之前，先介紹我國的對外貿易主管機構及從事對外貿易的主體。

一、外貿主管機構及出進口廠商

我國目前有三個主要外貿主管機構：經濟部國際貿易局，簡稱國貿局(BOFT─Board of Foreign Trade)，掌管輸出入許可；財政部所屬各海關專責貨品的進出口；而中央銀行(Central Bank of China)主管外匯有關事宜。凡我國內的公司、行號，其經營項目屬於出進口或買賣業務，且其資本額在新台幣伍佰萬元以上者，得憑(1)申請書及出進口廠商登記卡及(2)營利事業登記證影本，向國際貿易局申請登記爲出進口廠商。

出進口廠商前一年(曆年)之出進口實績達一定金額者，經濟部得予表揚爲績優廠商，並列入績優廠商名錄(詳附錄一：「出進口廠商登記管理辦法」)。

二、準備交易的階段

我國的貿易商經登記爲出進口廠商之後，不論是出口商或進口商，爲了開創業務、拓展外銷，或更新業績，都必須先做市場調查(Market Research)。因爲從市場調查的結果，出口商可以選擇可行性的外銷市場，而進口商則得到供應市場的情報。調查的方式，不外根據㈠自行前往實地調查或委託專業機構，實際調查所作的原始資料，或㈡利用他人蒐集整理之次級資料，如各項統計資料、各種年鑑、研究報告及刊物等，並配合自己的經營項目及目標，選擇具有發展潛力的市場。

　　選定市場之後，即可透過有關機構、同業、朋友、翻閱工商名錄、雜誌、刊物，或自行前往當地接洽，或參觀各項展覽會等，以便選擇股實可靠的交易對象。此際往來的書信，稱爲招攬業務函件(Letter or Circular Letter of Trade Proposal)。爲鞏固良好的往來關係，以及彼此長久的利益計，應委請往來銀行或徵信機構，進行徵信調查(Credit Inquiry)。調查的內容，最基本的乃三 C (=Character、Capital & Capacity)。Character 表達公司的誠實性(Integrity)、營運方針及公司的品格；Capital 表示公司的資產、資金及財務狀況(Financial Condition)；Capacity 則可以顯示公司的營運能力(Business Ability)、收益能力及發展的潛力。於是雙方存函建立業務往來關係。正規的或有相當規模的公司行號之間，尚需訂定一般貿易協定(General Terms and Conditions)，以爲日後交易的準則。

三、進行交易的階段

　　有業務往來的進口商與出口商之間，每筆交易，由詢價開始而終於履約。通常進口商爲欲購之貨品，去函查詢有關的規格、價格、付款及交貨等各項條件，而由出口商覆以報價。前者稱爲詢價函(Inquiry Letter)，後者爲報價函(Offer Letter)。報價乃報價者願意依照自己所開列之交易條件，與被報價人成立法律上有效之契約的意思表示。雖然報價的條件不只價格而已，但是價格卻是交易成立的關鍵要素，因此在我國的貿易界，習慣上稱之爲報價。報價必須經被報價人接受(Acceptance)，交易始能成交。若有部分條件不被接受，則雙方互相討價還價(Counter Offer)而至妥協爲止，否則交易無法成立時報價便消滅(Declined)。交易的成立，或以契約(Contract)，或以銷貨確認書(Sales Confirmation)確認之。比較積極的出口商，往往主動地報價，並且寄送目錄(Catalogue)、價目表 (Price List)或樣品(Sample)等以便爭取訂單。進口商方面，得憑出口商寄來的目錄、價目表或樣品等，挑選合意的項目，逐向出口商訂貨。有關訂貨的書信，往往是便於套打的格式，俗稱訂單 (Order Sheet)。我國的進出口廠商，在交易進行之階段，務必注意經手的貨品是否屬於准許類貨品。若屬於管制類貨品，則須先請示國貿局，以免成交後無法取得輸出或輸入許可證，而影響出口或進口，以致被控違約或背信。

四、履行交易的階段

買賣契約訂定之後，出口商和進口商的義務，分別為交貨和付款。買賣雙方約定以信用狀 (Letter of Credit) 為付款條件時，進口商應先履約，即開發信用狀；而出口商則於接獲信用狀後履約，即如約交貨。我國的進口廠商除免簽證類別的貨品之外，須先取得輸入許可證 (Import Permit)，再向中央銀行指定的外匯銀行 (Foreign Exchange Bank)，又稱指定銀行 (Appointed Bank)，辦理結匯並開發信用狀。我國的出口廠商，則於接獲國外開來的信用狀時，除免簽證項目的貨品之外，亦須申請輸出許可證 (Export Permit)，憑此辦理出口貨品報關後，依照信用狀的要求，準備必要的單據 (Required Documents)，開具匯票 (Drafts)，向往來的外匯銀行辦理出口押匯手續取得貨款。進口商則憑國外出口商的裝運通知 (Shipping Advice)，辦理投保 (CIF 報價者可免)，並準備貨款，於接獲開狀銀行的到單通知書，付款贖單，而經報關繳稅後提領貨品。

五、售後服務的階段

出口商欲獲取國外進口商不間斷的訂單，不但要注意國外市場的需求趨勢、講究優良的品質、保持交貨的準時，而且要追蹤銷售情形，並留意進口商的反應與意見，以鞏固銷路。遇對方來信抱怨 (Complaint) 或爭執 (Dispute) 時，應虛心自我檢討有無疏忽之處。若錯不在我，則舉證婉覆。反之，確係我方之錯，則函覆歉意並予實質的補償。此際雙方往來的書信，其內容即屬於貿易索賠 (Trade Claim)。倘若雙方無法自行解決，則改用仲裁(Arbitration)方式處理之。當糾紛的原因非屬進出口商雙方時，得按事故之肇因，分別向運輸公司和保險公司提出索賠要求。此乃屬於運輸索賠(Transportation Claim)和保險索賠(Insurance Claim)的範圍。

練習問題

一、試述我國主管外貿的三大機構。

二、欲從事對外貿易應如何着手？

三、試就對外貿易四個交易階段，摘要其主要的作業項目。

四、何謂三C？

五、下列各句是貿易書信的結尾用語(The Endings)。試指出其主旨(不必詳細翻譯)，並討論1.誰寫的？2.爲何而寫？

1. May we hear from you soon?

2. We hope you'll give us the opportunity to serve you again soon.

3. We suggest you drop in to discuss the matter with us. We shall be looking forward to seeing you.

4. As soon as our investigation is completed, we'll write to you again.

5. Please accept our best wishes for your continued success.

6. If there is anything we can do, please be sure to let us know.

7. We look forward to the opportunity of serving you again soon.

8. We hope this information will be helpful to you.

9. Will you please send us information. We will appreciate your reply soon.

10. Will you let us know by September 13 so that we can place our order promptly?

Trade English 第三章

貿易書信的結構

一、貿易書信的八大要素

目前的貿易書信在信箋上有八項必須記載的內容，稱為八大要素 (Eight Principal Elements)。按打字的順序依次為信頭、日期、受信人、稱呼、本文、結尾客套、簽署及識別記號。繕打時各項要素至少隔兩行為原則。

1.信頭(Heading or Letterhead)〔例3－1①〕

信頭佔信箋的上方醒目的地方，目的在於表示發文者的正式身份並提供回信有關的資料。信頭所列的項目不外發文公司或行號的名稱(Firm's Name)、地址(Address)、電話號碼(Telephone Number)及電報掛號(Cable Address)。租用郵政信箱者加列信箱號碼(P. O. Box Number)，裝備電報交換機(Telex)或傳真電報機(Fax)者分別標明電報交換號碼(Telex Number)或傳真號碼(Fax Number)。信頭通常印刷在信箋上方中間〔例3－1①〕，偶有佔左上方〔例3－2〕或右上方者〔例3－3〕。

2.日期(Date)〔例3－1②〕

日期乃寫信的日期(Letter Date)，以當地時間表示。表達方式分英國式與美國式兩種；英國式按日、月、年而美國式則照月、日、年之順序標出日期。由於英文的「日」屬於序數(Ordinal Number)，因此英國式的寫法將序數字尾兩個字母拼寫出來，但美國式則省略。月份務必正確地拼出不宜縮寫。年代之前，日、月或月、日之後英美兩式均標逗點。

1st January, 1998(英國式)

January 1, 1998(美國式)

至於全部使用阿拉伯數字的方式，如2/3/1998，則究竟是一九九八年二月三日抑或一九九八年三月二日，英美兩國人士的觀點不同，為免日後發生糾紛，儘量避免使用。

〔例 3 – 1 〕

① Solo Trading Co., Ltd.
Room 204, No. 71, Lane 501
Pei – An Road, Taipei, Taiwan 104
Republic of China

② October 25, 1998

③ American Market Corporation
Rm 2273, One World Trade Center
New York, New York 10048
U. S. A.

④ Gentlemen:

⑤ Thank you for your catalogue of the complete line of your automotive speaker systems and amplifiers.

Please quote us your best C&F prices for quantity shipments together with information on packing and delivery.

We look forward to your immediate response.

⑥ Yours faithfully,

⑦ Solo Trading Co., Ltd.

T. F. Li, Manager

⑧ TFL / myk

〔例3-2〕

ANZ Bank

Representative Office, 2nd Floor, 123 Nanking East Road,
Section 2, Taipei, Taiwan, R.O.C.

Telephone:508 4034   Telex:11894
Fax:508 3035

15 March 1988

Mr. Stuart T.F. Li
Senior Vice President &
General Manager Foreign Department,
Central Trust of China
49, Wu Chang Street, Section 1,
Taipei Taiwan 10006
Republic of China

〔例3-3〕

The First Canadian Bank

Bank of Montreal

Corporate & Government Banking

3210-3211, Connaught Centre,
1 Connaught Place, Central,
Hong Kong.
Telephone: 5-224182/8
Telex: 73731 MONT8 HX
Fax: 5-8104520

C. F. Davis
Senior Vice President

April 20, 1988

Central Trust of China
49, Wu Chang St., Sec. 1
Taipei, Taiwan 100
Republic of China

Attn. : Mr. Stuart T. F. Li
         Vice President &
         Deputy General Manager
---------Foreign_Department----

3.受信人（Addressee）〔例 3 －1 ③〕

貿易書信與我國書信最大的不同，在於信箋上一定要受信人的名稱和地址，故而又稱信內地址（Inside Address）。繕打時參照對方的信頭所載行號名稱地址，分繕四行至六行。第一行為公司行號名稱，第二行為門牌號碼及街道名，第三行為市鄉鎭名，第四行為國名。茲舉例如下：

> Southeast Trading Company, Inc.
>
> 2750 Monument Blvd.
>
> Concord, California 95204
>
> U. S. A.

若受信人係以人名為公司行號者，致函時宜冠 Mr. 的複數 Messrs.（＝Messieurs）如

> Messrs. Smith Brown & Co., Inc.
>
> 232, Madison Avenue
>
> New York, N. Y. 10016
>
> U. S. A.

致函對方之主管或有關人員時，於公司名稱之前加一行，繕打該主管或人員的稱呼、姓名及職稱，例如

> Mr. M. D. Swan, President
>
> Southeast Trading Company, Inc.
>
> 2750 Monument Blvd.
>
> Concord, California 95204
>
> U. S. A.

但是不知該主管姓名或不列姓名時則於職銜之前冠以定冠詞如：

> The Manager
>
> Southeast Trading Company, Inc.
>
> 2750 Monument Blvd.
>
> Concord, California 95204
>
> U. S. A.

4.稱呼(Salutation)〔例3－1④〕

稱呼相當於「敬啓者」，應與上面的受信人之性別、職稱對應。公司對公司的業務信函，通常使用之稱呼如下：

Dear Sirs, (英國式)

Gentlemen: (美國式)

英國式習慣用逗點而美國式則用冒號。當受信人欄出現個人姓名時，須注意其性別、職稱，加以適當的稱呼，例如：

Addressee	Salutation
Mr. M. D. Swan, President	Dear Sir,
The Manager	Dear Sir,
Dr. W. Richardson	Dear Dr. Richardson,
Miss Grace Dowson, Secretary	Dear Miss Dowson,
Mrs. Alice Kent, Section Chief	Dear Mrs. Kent, 或
	Dear Madam,

5.本文(Body of the Letter)〔例3－1⑤〕

本文乃貿易書信最主要的內容。本文的組織分開頭(The Opening Paragraph)、主文(The Main Body of the Letter)及結尾(The Closing Paragraph)。開頭和結尾各以一段2~5行爲宜。主文則按情節分段敍述，但各段落以不超過5~6行最爲適當。至於內容如何書寫，請參閱以後各章。

6.結尾客套(Complimentary Close)〔例3－1⑥〕

本文之後配合文意，選擇適當的客套作爲結束。常見的套句有

Yours faithfully,

Yours truly,

Yours very truly,

亦可變換使用如下：

> Faithfully yours,
>
> Truly yours,
>
> Very truly yours,

對於關係比較密切者可用

> Yours sincerely,
>
> Sincerely yours,
>
> Cordially yours, 或
>
> Yours cordially,

欲表達敬意者可用

> Yours respectfully, 或
>
> Respectfully yours,

7.簽署 (Signature)〔例 3 −1 ⑦〕

貿易書信為證實其信函的內容，由主事者或有權簽字人簽署，並加打姓名及頭銜以示負文責，例如：

Roger R Meadows

Roger R. Meadows

President

由公司授權各部門主管負責簽署的方式如下：

SOUTHERN SUPPLY COMPANY

Adam D. Lassiter

Adam D. Lassiter

Credit Manager

代理簽署的方式如下：

for ELLIOT ENTERPRISES, INC.   BARKER'S BOOTERY

Lawrence M. Grant   by *Wanda Wexler*

8.識別記號(Identification Marks)〔例3－1⑧〕

發文者為表明分層負責起見,於簽署欄之下隔兩行左下角處,繕打簽署人和打字員的姓名字首。前者負文責用大寫字首,後者負打字校對之責,以小寫字首表示,因此又稱識別字首(Identification Initials)如 MDS / gd 或 MDS: gd。

二、貿易書信的附帶要素

貿易書信除了以上八項不可或缺的要素之外,得視實際需要酌予列入信箋者有案號、特定受信人、主旨、附件欄、副本抄送標識及附註欄等六項附帶要素(Additional Elements)。

1.案號(Reference Number)〔例3－4①〕

為便於處理、照會及歸檔,貿易書信往往在左上角編列案號。來函有編號者回信時應予先列對方案號,再打出我方案號如:

Your Ref. No. 450

Our Ref EX －36

2.特定受信人(Particular Address)〔例3－4②〕

貿易書信原則上以公司行號為受信人,但是有必要先請其主管或特定人過目關照時,得於受信人和稱呼之間以 Attention: 表達,此際稱呼不對該個人而仍然以對公司之稱呼為稱呼。例如:

Attention: Mr. J. William, Vice President

Attention: Export Manager

Attention: Import Department

〔例 3 – 4 〕

AMERICAN MARKET CORPORATION

Rm 2273, One World Trade Center
New York, New York 10048

November 2, 1998

① Your Ref.
Our Ref. EX – 035

Solo Trading Co., Ltd.
Rm 204, No. 71, Lane 501
Pei – An Rd, Taipei, Taiwan 104
Republic of China

② Attention: Mr. Stuart Li, Manager

Gentlemen:

③   Re: our quotation

In response to your letter dated October 25, 1997, we have quoted the lowest prices C&F Keelung for our automotive speaker systems and amplifiers.

Enclosed is our quotation with information on packing and delivery.

Your patronage will be appreciated.

Yours sincerely,

AMERICAN MARKET CORPORATION

Adam E. Wood
Export Manager

AEW / egm

④ Encl.

⑤ c. c. Wang & Brothers Co., Ltd., New York

⑥ p. s. Banking charges are for your account.

3.主旨(Subject)〔例3-4③〕

為使受信人易於掌握本文所提案情，得將本文的主旨以片語的方式繕打於本文之上方。常見的表達方式如下：

(1)　　Re: Your Hi-Fi Speaker System

(2)　　Subject: Inspection Certificate

(3)　　Our Order No. TT-234

4.附件欄(Enclosure Remarks)〔例3-4④〕

隨函附送文件時，為提醒受信人起見，於左下角用 Enc. 或 Encl. 表示。至於附件的名稱與數量已於本文中提及者可以省略。

5.副本抄送標識(Carbon Copy Notation)〔例3-4⑤〕

函件內容有照會或抄送第三者之必要時，以 c. c. 表示並打出收受者名稱如：c. c. B. L. Drew & Co., New York

6.附註欄(Postscript)〔例3-4⑥〕

本文未提或漏提而必須補充之事項，得以 P. S. 標出應補充的內容。惟除非有特殊意義，否則寧可重繕本文以免予人不良印象。

三、貿易書信的體裁

書寫或繕打受信人、本文及簽署等三項要素的方式有二：(1)縮退型(Indented Form)，即第二行以下每一行的開頭均較前一行縮退者及(2)齊頭型(Block Form)，即第二行以下每一行的開頭均與前一行看齊者。茲分別舉例如下：〔例3-5〕。

由此衍生四種貿易書信的體裁(Styles of Business Letter)

〔例 3 – 5 〕

(一)　Inside Address　受信人
(1) Indented Form:　American Market Corporation
Rm 2273,　One World Trade Center
New York,　New York 10048
U. S. A.

(2) Block Form:　American Market Corporation
Rm 2273,　One World Trade Center
New York,　New York 10048
U. S. A.

(二)　Body　本文
(1) Indented Form:　Thank you for your catalogue of the complete line of your automotive speaker systems and amplifiers.

Please quote us your best C & F prices for quantity shipments together with information on packing and delivery.

We look forward to your immediate response.

(2) Block Form:　Thank you for your catalogue of the complete line of your automotive speaker systems and amplifiers.

Please quote us your best C & F prices for quanity shipments together with information on packing and delivery.

We look forward to your immediate response.

(三)　Signature　簽署
(1) Indented Form: Solo Trading Co., Ltd.

T. F. Li,　Manager
(2) Block Form:　Solo Trading Co., Ltd.

T. F. Li,　Manager

1 縮退體裁 (Indented Style)

受信人、本文和簽署等三項要素均採用縮退型的貿易書信，其體裁稱為縮退體裁〔例 3－6〕。

〔例 3－6〕

January 12, 1998.

Mr. Donald J. Roberts.
　　922, Hillcrest Avenue,
　　　Jefferson, Illinois.

Dear Mr. Roberts:

　　This letter is an example of the indented or steeped – in form. The date line is arranged in the upper right – hand portion of the sheet. The first line of the inside address sets the left – hand margin of the letter. Each additional line of the inside address is uniformly indented either three or five spaces more than the preceding line.

　　Paragraph beginnings are usually indented the same amount as the last line of the inside address, although they may coincide with the indention of the next to the last line of the inside address. The complimentary close ordinarily aligns with the date line. The signature is written two or three spaces to the right of the point at which the complimentary close begins.

　　The indented form is used by a number of conservative organizations that prefer the longestablished method to its newer variations. This form is acceptable, however, for the correspondence of any business or professional man.

Sincerely yours,
(Signature)
Leon E. Harris
Account Executive

2.齊頭體裁(Block Style)

貿易書信的受信人、本文和簽署均採用齊頭型者，其體裁稱為齊頭體裁〔例3－7〕。

〔例3－7〕

March 5, 1998

Mr. J. C. Cummings
347 East Oak Street
Council Bluffs 10, Iowa

Dear Mr. Cummings:

This letter illustrates the block form of letter dress, which has become one of the most widely used methods of arranging letters.

It takes its name from the fact that the inside address, the salutation, and the paragraphs of the letter itself are arranged in blocks without indention. The block form offers two distinct advantages; it saves stenographic time and reduces the number of margins. Its wide acceptance at the present time offers assurance that the letter arranged in block form is correct and modern.

If you desire your letters to be attractive in appearance, modern, and economical with regard to stenographic time, I heartily recommend the block form as the most suitable for the needs of your office.

Sincerely yours,
(Signature)
Geraldine A. Fisher
Correspondence Supervisor

3.半齊頭體裁(Semiblock Style)

　　受信人和簽署採用齊頭型而本文仍採用縮退型的貿易書信，其體裁稱為半齊頭體裁〔例3-8〕。

〔例3-8〕

March 5, 1998

Mr. Robert C. Vanderlyn
2202 Middlebury Road
Winchester 4, Maine

Dear Mr. Vanderlyn:

　　I appreciate your interest in my reasons for recommending the type of letter arrangement which our company uses in its correspondence.

　　After careful consideration, I recommended the semiblock form as the most effective for our company. This recommendation was based on my belief that this form combined most of the advantages of the block and the indented forms.

　　The block arrangement of the inside address appeals to me as symmetrical and economical of secretarial time; furthermore, open punctuation is modern and efficient. Perhaps it is no more than a whim on my part, but I prefer to have the paragraphs of the actual message indented as they are in books, newspapers, and magazines.

　　The semiblock form meets all these requirements; it has proved effective and is well liked by our staff of correspondents and secretaries after six years of use.

Sincerely yours,
(Signature)
John H. Porter
Correspondence Supervisor

4.全齊頭體裁(Fullblock Style)

　　將日期、結尾客套及簽署三項要素移至左邊與受信人的位置看齊，並且採用齊頭型的貿易書信，其體裁稱為全齊頭體裁〔例3-9〕。

〔例3-9〕

March 5, 1998

Mr. Donald E. Woodbury
3126 Westview Road
Seattle 5, Washington

Dear Mr. Woodbury:

Your comments about the form of our letters interested me greatly. As you pointed out, letters do reflect the personality of the firm which sends them, and that fact played a large part in our decision to adopt the complete or, as it is sometimes called, the full-block form.

As management consultants, we felt that our letters should exemplify the same standards of efficiency and the modern methods we advocate in industry. For that reason, we saw no sound reason for retaining a letter form which requires changes of margins and unnecessary stenographic time.

The salient features of the full-block form are illustrated in the letter. You will be interested to know that we have received a number of favorable comments about our letter form and that our Stenographic Department likes it very much.

Sincerely yours,
(Signature)
E. J. Baumgartner
Partner

四、貿易書信的標點型式

　　除了信頭及本文之外，繕打貿易書信的各項要素時，若每行終了均加標點者，稱爲關閉式標點（Closed Punctuation），而每行終了並不加標點者，則爲開放式標點（Open Punctuation）。但每行末了一個字係縮寫時，簡寫符號仍應予保留，茲舉例如下。

Closed Punctuation:

　　　　Solo Trading Co., Ltd.,

　　　　Rm. 204, No. 71, Lane 501,

　　　　Pei－An Rd., Taipei, Taiwan 104,

　　　　Republic of China.

Open Punctuation:

　　　　Solo Trading Co., Ltd.

　　　　Rm. 204, No. 71, Lane 501

　　　　Pei－An Rd., Taipei, Taiwan 104

　　　　Republic of China

　　從使用標點的觀點而言，目前的貿易書信可歸類爲以下三種標點型式（Punctuation Patterns）。

1.關閉式標點型式 (Closed Punctuation Patterns)

凡信頭及本文以外，其餘各項要素均採關閉式標點者屬之，如〔例 3−10〕。

〔例 3−10〕

HEADING

October 10, 1998.

Solo Trading Co., Ltd.,
Rm. 204, No. 71, Lane 501,
Pei − An Rd., Taipei, Taiwan 104,
Republic of China.

Gentlemen:

    With reference to your inquiry dated October 1, 1998, we are enclosing our catalog on speaker systems and amplifiers.

    Your patronage will be appreciated.

Yours truly,

(SIGNATURE)
L. M. Grant,
Manager.

LMG / gib
Enc.

2.混合式標點型式 (Mixed Punctuation Patterns)

　　信頭及本文除外，其餘各要素中，只有稱呼和結尾客套採用關閉式標點，其餘則採開放式標點者，如〔例3－11〕。

〔例3－11〕

```
                         HEADING


                                      October 10, 1998


Solo Trading Co., Ltd.
Rm. 204, No. 71, Lane 501
Pei－An Rd., Taipei, Taiwan 104
Republic of China

Gentlemen:

With reference to your inquiry dated October 1, 1998, we are enclosing our
catalog on speaker systems and amplifiers.

Your patronage will be appreciated.


                                      Yours truly,

                                      (SIGNATURE)
                                      L. M. Grant
                                      Manager


LMG / gib
Enc.
```

3.開放式標點型式(Open Punctuation Patterns)

　　信頭及本文之外,其餘各項要素均採開放式標點者屬之。這是最新的標點型式。近年來不但在美國的軍事機構通行,一般公司行號亦逐漸採用,如〔例3-12〕。

〔例3-12〕

H E A D I N G

October 10, 1998

Solo Trading Co., Ltd.
Rm. 204, No. 71, Lane 501
Pei-An Rd., Taipei, Taiwan 104
Republic of China

Gentlemen

With reference to your inquiry dated October 1, 1998, we are enclosing our catalog on speaker systems and amplifiers.

Your patronage will be appeciated.

Yours truly

(SIGNATURE)
L. M. Grant
Manager

LMG /gib
Enc.

五、信封(Envelope)

　　貿易書信所用的信封,有以下四項繕打內容〔例3-13〕。

〔**例3-13**〕

② Return Address
(Sender's Name & Address)

Stamp

① Mail Address
　(Addressee's Name & Address)

④ Remarks
③ Mail directions

1.郵遞地址(Mail Address)

　　將信箋上的受信人名稱地址,以同樣的體裁、標點型式繕打於信封上〔例3-13①〕。

2.回信地址(Return Address)

　　回信地址係發文者的名稱地址,通常印刷在信封的左上角〔例3-13②〕,有時加打" If undelivered please return to "以便郵件無法投遞時退回之用,故稱之為回信地址。

3.郵寄指示(Mail Directions)

　　若欲以航空交寄郵件者,應於信封上左下角〔例3-13③〕標出 AIR MAIL, By Air Mail, VIA AIR MAIL, 或 PAR AVION 等字樣。但是使用印便的航空信封則免打。

4.附註欄(Remarks)

常見的附註如下：
Printed Matter(印刷品)
Sample of No Value(免費樣品)
Confidential(密件) Private, Personal, (親啓)
Registered(掛號)

練習問題

一、信頭是甚麼？有何作用？

二、日期有幾種寫法？試述其重要性。

三、試述繕打 Inside Address 應注意事項。

四、試說明 Salutation 和 Complimentary Close。

五、試述如何安排本文。

六、試述簽署部分如何繕打。

七、貿易書信的附帶要素有那幾項？

八、試述貿易書信的體裁。

九、試述貿易書信的標點型式。

十、信封上經常繕打的項目為何？

Trade 第四章 English

貿易書信的撰寫要訣

近代的貿易書信講究效率，即以最精簡的語句達到傳遞信息和迅速獲得回音的目的。撰稿的要訣有六：

一、措辭要正確（Be Correct）

正確是撰稿的基本原則。正確的意思表達靠意義正確的字眼，配以正確的文法和語法，並且用正確的拼音、標點和大小寫書寫。名詞的「數」往往左右字義，尤其是撰寫貿易書信，一字之差可能前功盡棄。茲舉數例以供參考。

1.不用複數的名詞

(1) Thank you for your various informations.　（誤）
Thank you for your various information.　（正）
謝謝貴公司的各項資料。

(2) We shall send 1,000 dozens of hosiery.　（誤）
We shall send 1,000 dozen of hosiery.　（正）
本公司將寄壹千打襪子。

(3) We mailed them a catalog of general merchandises. （誤）
We mailed them a catalog of general merchandise.　（正）
本公司曾郵寄該公司一份雜貨目錄。

(4) They inquired about our machineries.　（誤）
They inquired about our machinery.　（正）
該公司洽詢本公司的機器。

2.不用單數的名詞

下列的名詞其單數與複數具有不同意義，貿易上限用複數者。

instructions	指示	instuction	教導
contents	內容	content	滿足

conditions	條件	condition	情況
terms	條件	term	學期
arrangements	籌備	arrangement	安排，佈置

3.數量(Quantity)的詞語當形容詞時，不可使用複數

2－meters－wide board	(誤)
2－meter－wide board	(正)兩公尺寬的板
5－feet－long boat	(誤)
5－foot－long boat	(正)五英尺長的遊艇
12－stories building	(誤)
12－story building	(正)十二層的大廈
6－hours tour	(誤)
6－hour tour	(正)六小時的旅遊
30－minutes drive	(誤)
30－minute drive	(正)三十分鐘的車程
10－tons truck	(誤)
10－ton truck	(正)十噸重的卡車
20－pounds drum	(誤)
20－pound drum	(正)二十磅重的圓桶
25－square－meters carpet	(誤)
25－square－meter carpet	(正)二十五平方公尺的地氈

又 staff 表示職員時之用法如下：

He is a staff of our company.　　　(誤)

He is a staff member of our company. (正)

他是本公司的職員。

They are staffs of our company.　　(誤)

They are staff members of our company. (正)

他們是本公司的職員。

動詞和介系詞是意思表達的經緯。這兩種詞類均有其慣用的語法。茲舉數例如下：

recommend, suggest

　　We recommend you to become a member.　　（誤）

　　We recommend (that) you become a member.　（正）
　　本公司推薦貴公司當會員。

　　He suggested you to go there.　　（誤）

　　He suggested you go there.　　　（正）
　　他提議由你前往。

permit, prohibit

　　It is not permitted to you to do so.　　（誤）

　　It is not permitted for you to do so.　　（正）

　　=You are not permitted to do so.
　　不容貴公司有此種行為。

　　It is prohibited to all companies to do business with that party.
　　（誤）

　　It is prohibited for all companies to do business with that party.
　　（正）

　　=All companies are prohibited to do business with that party.
　　不許各公司與該單位進行交易。

reach, visit, ask, contact

　　The goods reached to the warehouse yesterday.　　（誤）

　　The goods reached the warehouse yesterday.　　（正）
　　貨品於昨日進倉。

　　They will visit to your company about October.　　（誤）

　　They will visit your company about October.　　（正）
　　他們將於十月間拜訪貴公司。

inquire

We inquired the company about the merchandise.　　（誤）

We inquired of the company about the merchandise.　　（正）

＝We inquired the merchandise of the company.

我們向該公司洽詢貨品。

exchange

She exchanged her apple with his peach.　（誤）

She exchanged her apple for his peach.　　（正）

她以蘋果交換他的梨子。

discuss, explain

We have discussed about the matter with them.　　（誤）

We have discussed the matter with them.　　（正）

我們跟該公司討論過這件事。

Let me explain about the matter.　　（誤）

Let me explain the matter.　　（正）

容我解釋。

exceed

It exceeds above 100 dollars.　　（誤）

It exceeds 100 dollars.　　（正）

金額超出一百元。

furnish, provide, supply

We furnished them our business machines.　　（誤）

We furnished them with our business machines.　　（正）

We provided them with our business machines.　　（正）

We supplied them with our business machines.　　（正）

本公司以事務機器供應該公司。

attach

It is attached with the machine.　（誤）

It is attached to the machine.　　（正）

那是機器的附件。

enter

> We entered into the hall.　　（誤）
>
> We entered the hall.　　（正）
>
> 我們進入了大廳。
>
> We have entered business relations with them.　　（誤）
>
> We have entered into business relations with them.　（正）
>
> 本公司和該公司已建立業務往來關係。

increase

> Profits will be increased.　　（誤）
>
> Profits will increase.　　（正）
>
> 利潤將增加。

succeed

> We have succeeded the contract.　　（誤）
>
> We have succeeded in the contract.　　（正）
>
> 我們完成了合約。

until, by

> They were making the package by eight o'clock.　　（誤）
>
> They were making the package until eight o'clock.　（正）
>
> 他們打包到八點鐘。
>
> We shall open an L / C until December 20, 1998.　　（誤）
>
> We shall open an L / C by December 20, 1998.　　（正）
>
> 本公司將於十二月二十日以前開出信用狀。

about

> about at noon　　　　（誤）
>
> about noon　　　　（正）午間
>
> about on June 3rd　　（誤）
>
> about June 3rd　　　　（正）六月三日左右

about in October　（誤）

about October　　（正）十月份

The cargo will arrive at Keelung about July 5 or 6.　（誤）

The cargo will arrive at Keelung on July 5 or 6.　　（正）

貨物可望於七月五日或六日運抵基隆。

其他尚需注意者如下：

convenient

If you are convenient, may I see you？　　（誤）

If it is convenient to you, may I see you？　　（正）

＝Will it be convenient for you to see me？

倘若方便可否晤談。

following

The followings are our opinions：（誤）

The following are our opinions：（正）

本公司的建議如下：

direct, directly

I shall go directly to New York.　（誤）

I shall go direct to New York.　（正）

我將逕赴紐約。

acceptable

Will it be acceptable for you？　（誤）

Will it be acceptable to you？　（正）

不知貴公司是否接受？

二、語意要具體（Be Concrete）

具體的原則建立於清楚的基礎上。合乎正確的書信若使用意義不清楚的字眼，則其敘述未必能符合原意。例如 Please refer to our previous letter sent to you. 此句正確地表達了「請參閱前函」的意思，但是 " previous " 係指先前的，因此受信人無法確定係指那一封，此乃不清楚的字眼。爲表達具體的語意，宜將去函日期清楚地列出，例如 Please refer to our letter dated May 20, 1998. 則對方一眼即知。

同樣地 We have received your communication. 只能表達了「獲悉」或「敬悉」。至於獲悉甚麼，敬悉甚麼，由於 Communication 一詞籠統地含蓋所有連繫的工具，因此未交代清楚。若來文係書信則用 your letter，電文則用 your cable 或 telex 或 fax，洽詢函則以 your inquiry 來表達更爲清楚而又具體。

在信尾常見的「敬請惠示……」若以 Please advise us of……表達，則雖然沒有錯但是不切實際。因爲 advise 有勸告和通知兩種含義，所以要求對方通知時，不如改用 Please inform us of…豈不更具體。

再者，We are sending our sample in a week. 此句究竟指「一週內即寄樣品」，抑或「過一週即寄樣品」令人費解；蓋 in 可指時間之過程和時限。因此欲表達週內則宜改用 within a week 較妥當。

當國外進口商來函洽購而庫存不多時，如何覆函呢？有人寫 Fortunately we have a very limited quantity of the goods. 以爲可以表達「幸虧庫存尙有微量」。殊不知按英文的語法此句只能解釋「本公司慶幸僅有一些存貨」。如此可能引起對方誤會，厚道的生意人豈可奇貨可居。這又是語意不清作祟。原意應爲 Fortunately we have the goods, although very limited in quantity.

茲有一段覆國外訂單的書信。

We have received your letter dated September 14, 1998 and are pleased to inform you that the goods ordered by you will be forwarded to you as soon as we are in a position to supply them.

大意是九月十四日來函敬悉，惠訂之貨品可望有現貨時即奉寄。本文雖然平易但是能否如期交貨，未作肯定的答覆，買方可能不悅而取消訂單。應積極地將具體的處理方式函覆對方取得諒解才是上策，例如

Thank you for your letter dated September 14, 1998 together with the enclosed order sheet.

Please be informed that our manufacturer has promised us a fresh supply in about ten days so we can assure you that the whole quantity will be shipped out to you by the end of this month.

再看下一句

His report is not worthwhile. (他的報告毫無可取)

句中的 worthwhile 是含蓄的字眼，無法讓讀者了解撰稿者的原意。若改以客觀而具體的敘述如

His report doesn't give the information I asked for. (他的報告不符本人的要求)則客觀的敘述更顯示具體的事實。

同樣地 His lecture was very worthwhile. (他的演講值得一聽)是主觀的敘述，若改為 His lecture was educational, well prepared, worth two hours of your time. (他的演講內容充實具有啓發性，不枉費你兩個鐘頭)則不但客觀而且更能傳達演講的盛況。

三、敘述要簡潔 (Be Concise)

簡潔的敘述能讓讀者在最短時間內把握重點，引起他的興趣而達到傳遞信息的目的。簡潔的第一個步驟是刪去囉嗦和多餘的字。

May we hear from you at an early date? (尚請盡速賜覆)是語意清楚的句子，但是 at an early date 在現代係屬於囉嗦浪費時間的字眼。若以 soon 代之而改為 May we hear from you soon? 則較簡潔而更平易近人了。

We must get to the basic fundamentals of the problem. (我們必須認清此一問題的基本原則)這是意思表示非常清楚的句子。但是 fundamental 的意義為基本原則，因此 basic 似乎是多餘的，刪去則更為簡潔。

類似的表達如：

exactly identical　　完全一樣

quite unique　　　　獨特的

past experience　　經驗

由於 identical 即為完全一樣 (exactly alike), unique 即唯一的 (the only one of its sort)，而 experience 當然是由過去的事情累積得來的經驗。因此 exactly, quite, past 似乎又是多餘的。

　　簡潔的第二個步驟是語調必須合乎潮流。古板拘束、慢條斯理、拐彎抹角的敘述已不適於現代的社會。

　　像 We are in receipt of your check. (支票收妥) 是五十年前的貿易英文，現代人用 We received with thanks your check. 或更簡潔的 Thank you for your check.

　　至於 We have duly noted your letter of May 30, 1998. (五月三十日大函敬悉) 更使現代人摸不着頭腦，應改為 We have read your letter of May 30, 1998.

　　以下數例乃不合時代的語句，宜分別使用右邊各句：

We acknowledge receipt of…　　(……敬悉)

　　　　　　　　　　　Thank you for…

　　　　　　　　　　　We received your…

Attached please find…　　　　(隨函檢送……)

　　　　　　　　　　　We are enclosing…

You will herewith find…　　　(隨函檢奉……)

　　　　　　　　　　　Enclosed is…

Due to the fact that…　　　　(鑒於……)

　　　　　　　　　　　Because…

　　　　　　　　　　　Since…

In view of the fact that…　　(由於……)

Owing to the fact that…　　　(由於……)

　　　　　　　　　　　Because…

	Since···
Thank you in advance···	(謹謝······)
	We shall appreciate···
Will you be kind enough to	(敬請······)
	Please···
	Will you···
We take the liberty to send you···	(茲檢送······)
It is our great pleasure to send you···	
We have the pleasure of sending you···	
We take pleasure in sending you···	
	We are sending···
	We are pleased to send···
	We send you···
In compliance with your request···	(茲遵囑······)
	As requested···
at the present time···	(現今······)
at the present moment···	
	now
at your earliest convenience···	(務請儘速······)
at the earliest possible time···	
	soon

四、文章要有力量 (Be Powerful)

動態的敘述能提高讀者的興趣，使他盯住你的敘述而欲罷不能。

例如

There was a week of discussion of commodity prices by the delegates.

和 The delegates discussed commodity prices for a week.

同樣地報導「與會代表曾就物價問題商討了一週」，但是前者用狀態動詞 be 因而敘述平平無奇；後者則使用動態動詞 discuss 因而敘述比較活潑令人注目。再看下面的廣告：

A feature of the OK Computer is its capacity to bring about a solution of a mathematical problem in one —tenth of a second.

（歐開型電腦具有特殊性能，可於十分之一秒內解答數學方面的問題。）

本句雖然簡要地敘述該電腦的性能，但是以狀態動詞爲主動詞，無形中削弱了說服的力量。本句隱藏三個動態動詞—這種非爲句中的主動詞，但是可以使用爲動詞者，稱爲隱性動詞(Hidden Verbs)—即 feature, bring about 及 solve。利用這三個隱性動詞可以修改本敘述如後：

（1）若以 to feature 爲主動詞，則本句可以改寫爲

The OK Computer features a capacity to bring about a solution of a mathematical problem in one —tenth of a second.

（2）若以 to bring about 爲主動詞，則可改寫爲

The OK Computer can bring about the solution of a mathematical problem in one —tenth of a second.

（3）若以 to solve 爲主動詞，則可改寫爲

The OK Computer can solve a mathematical problem in one —tenth of a second.

以上經修改後的句子，均比原來的敘述更具說服的力量，尤其以第(3)句爲最。

五、文章要有組織(Be Organized)

個別的敘述雖然合乎具體、簡潔以及有力的要訣，但是未經過組織的文章不過是片斷的敘述，無法讓讀者了解主旨何在。文章的組織通常以主句(Main Statement)表達主旨(Main Idea)，修飾語(Modifiers)表達說明部

分。修飾語按說明的份量由重而輕分別以子句、片語及單字表達。下面的
敘述由三句構成：

> We want to bring existing statistics up to date. To do this, we will be-
> gin a survey of truck transportation. June will be the starting time.

由於毫無組織令人不知所云。若能提出主旨配以適當的說明即可成為有條
理的敘述。例如：

> To bring existing statistics up to date, we will begin a survey of truck
> transportation in June.

惟組織文章尚須節制修飾以免喧賓奪主而埋沒了主旨。例如

> We found the missing documents, which were located under a pile of
> newspaper.

(我們從一堆報紙裏尋回遺失的文件)

本句中主旨是尋回文件，其份量和說明該項文件的部分齊等，因此讀起來
平淡毫無感受。若用片語修飾，以減輕說明部分的份量如：

> We found the missing documents under a pile of newspaper.

則不但令人共鳴尋回的高興，而且知道原來藏在報堆裏之後會不覺一笑。
至於

> Auto production and sales increased sharply last November and this
> occurred particularly in the Detroit Area.

(十一月份的汽車產銷均增，尤其以底特律地區為最)

此句將兩個應有輕重之分的子句，用對等連接詞連接後，令人不知其主旨
何在。若以前一句為主旨時可改為

> Auto production and sales particularly in the Detroit Area increased
> sharply last November.

如此一來主題格外分明令人易懂。

六、語氣要中肯(Be Courteous)

　　書信往來如同待人接物，自然的語調、友善和謙虛必能予人好感而得到回音。所謂自然的語調，乃是讀者所熟悉而聽慣的語調，談話式的語調便是其中之一。

像 It's operational deficiencies were attributed by the agency to a lack of personal resulting from budget limitations.

（該機構工作效率之所以欠佳，據代理商認爲乃限於預算、員額不足所致。）

句中不少鑽牛角尖、落伍的措辭，令人乏味易生反感。改寫爲

The agency said that it could not do a good job because it did not have enough money to hire enough people.

則因使用讀者熟悉的語句，讀起來甚爲自然。

　　維護同業間業務的往來，必須建立於互惠的原則上。同樣，互助及建議性的語氣，乃友善和關心的表現，必然能獲得回報。像

We are sorry that we've kept you waiting but it's taken us a little more time than we expected to process your application. We'll have it completed by next Friday.

（台端的申請案，因需費時較久，請見諒。惟於下週五之前當可辦妥，惠請查照）

雖然申請事宜尚未辦妥，但是語氣友善而且肯定，原申請人當可見諒無疑。相對地

We are obliged to inform you that the completion of processing of your application has been delayed until the end of next week.

則有點官腔，毫無關懷之情。申請人接到通知後可能嘀咕甚至引起反感。

　　常說 Please 和 Thank you 是謙虛的具體表現。如

Submit your check for US$20.00 not later than December 10, 1998.

要求於十二月十日之前惠寄美金支票二十元一紙。目前公營機構都常用「請」字，貿易界的同業諒不應有如此霸道者。最好改寫為

May we please have your check for US$20. 00 by December 10, 1998 ?

或 Please send your check for US$20. 00 by December 10, 1998.

　　謙虛有禮的另外一種表達方式，是在文中多提對方。例如修函問候對方過去兩週所遭遇的困難而寫

　　It is hoped that the difficulties of the past two weeks have been over-
　　come.

則句中顯然以事為主題，語氣似乎無關痛癢。若能以當事人為主題，例如

We hoped that you have overcome your difficulties of the past two weeks.

則不但關心之情洋溢紙上而且更富人情味。

練習問題

一、貿易書信撰寫的要訣有六，其目的何在？

二、試述措辭如何正確(Be Correct)？

三、試述語意如何具體(Be Concrete)？

四、試述敘述如何簡潔(Be Concise)？

五、試述文章要有力量(Be Powerful)的原則。

六、試述文章要有組織(Be Organized)的原則。

七、何謂 You Attitude？與語意要中肯(Be Courteous)有何關聯？

八、試將 impression, action, inform 三字填入適當的空格。

　　The effective business letter can do three things:

　　1. It can 　　　　the reader.

　　2. It can get 　　　　from the reader.

　　3. It can make a good 　　　　on the reader.

九、Complete each sentence with the most specific word or phrase.

　　1. We will be happy to send you a check ___ .

a. in due course

b. as soon as possible

c. by August 15

2. The court asked the taxpayer to ___ his taxes.

    a. pay

    b. arrange

    c. take care of

3. The manager noticed that the lights were out, so he ___ that the workers had gone home.

    a. thought

    b. believed

    c. concluded

4. We wished to obtain further information, so we ___ the assignee.

    a. communicated

    b. contacted

    c. telegraphed

5. The trainee ___ questions promptly and clearly.

    a. responded to

    b. answered

    c. reached to

6. We will be glad to take care of this ___ for you.

    a. matter

    b. problem

    c. debt

7. The profits from the three subsidiaries ___ US$545, 356. 00.

    a. were

    b. came to

    c. totaled

十、Match the simple short word with the inflated word which has exactly the same meaning.

1. approximately
2. subsequently
3. anticipated
4. causative factor
5. optimum
6. inaugurate
7. finalize
8. manifest
9. replicate
10. recapitulate

a. finish
b. repeat
c. best
d. copy
e. begin
f. about
g. cause
h. expected
i. show
j. later

Trade 第五章 English

交易的準備

一、尋找交易對象

選擇交易的對象，從了解對方的市場狀況着手。假使對方的市場頗具潛力，則交易的對象縱然是二流的廠商，我方仍可望獲得長久的利益；反之，對方的市場已呈疲態而無甚展望時，對方雖然是一流的廠商，恐已無利可圖，必然枉費所投注的金錢與精神。衡量國外市場，尚應摒棄主觀的眼光和跟進的作風。蓋前者獨斷的想法，往往成為商品打進生活習慣和價值觀念均與我方不甚相同之國外市場的障礙；後者不但容易造成一窩蜂的現象，而且會引起對方國家採取設限的危險，兩者均對開拓市場有害無利。

日後交易的對象，應該是從適當的市場或地區內選擇幾家廠商加以過濾後選定者。選擇的方式，可以分為積極的和消極的；前者乃自行接洽，而後者則委請有關機構推薦適當的對象。自行接洽的方法有：(1)翻閱工商名錄和廣告；(2)參觀或參展國內外之貿易展覽；及(3)前往國外。

工商名錄(Directory)有按國家地區或按進出口商分別編印者，其中以英國的凱利公司出版的凱利廠商名錄(Kelly's Directory of Manufacturers & Merchants)最為著名。貿易廣告(Trade or Business Advertisement)則散見於國內的各大報、英文報及相關行業的專刊及雜誌等。在國內舉辦的貿易展覽會(Trade Exhibition or Shows)，一則可於展示期間前往參觀國外廠商的商品，並與之接洽；二則可以參展，以便廣招國外廠商前來參觀並接洽。在國外舉行的貿易展覽會，可以申請出國參觀或透過有關機構申請參展。至於前往國外，則可自行出國或隨工商考察團出國前往接洽。

採取委請推薦方式時，在國內有下列各機構值得去函委請推薦適當的廠商：

1. 經濟部國際貿易局(BOFT; Board of Foreign Trade, MOE)
2. 中華民國對外貿易發展協會(CETDC; China External Trade Development Council)

3.台北市商業會(Taipei Chamber of Commerce)

4.高雄市商業會(Kaohsiung Chamber of Commerce)

5.台灣省商業會(Taiwan Chamber of Commerce)

6.有業務往來之各家外匯銀行(Foreign Exchange Banks)

7.有業務往來之外商銀行(Foreign Banks in Taiwan)

8.各國駐華領事館或大使館(Consulate or Embassy in Taiwan)

委請國際貿易局、對外貿易發展協會、各商業會及有往來的外匯銀行等機構推薦廠商時,當然以中文函;但委請外商銀行及各國駐華領事館或大使館推薦時宜用英文函。國外方面,各大都市均有商會(Chamber of Commerce)的組織。透過各地商會推薦適當的貿易對象必須借助於貿易書信。

(一)函請商會推薦進口商

台北市大直的實踐貿易公司(Shih Chien Trading Co., Ltd.),經營木製傢俱(Wooden Furniture)、家庭用品(Household Articles)及各項雜貨(Sundry)。近欲拓展中東外銷市場,選定了沙烏地阿拉伯,於是致函吉達的商會委請介紹殷實可靠的進口商〔例5-1〕。

致函商會,通常以總幹事(The Secretary)為受信人(Addressee),如同致函銀行時以經理(The Manager)為受信人一樣。

書寫貿易書信的動機,可大別為主動的去信和被動的回信;前者多為通知或要求的函件,而後者必然是對於前者的覆函。

本函屬於主動而帶有要求的去信。於開頭(The Beginning)開門見山地敘述去信的目的,接着自我介紹並提備咨人(reference),而以道謝的詞句結束。

第一段以動態動詞 desire 打開話題,而以 grateful 點到寫信的目的。

We desire…and would be grateful if you…

(為……煩請……是幸)

〔例 5 – 1〕

The Secretary
The Chamber of Commerce of Jedda
Jedda, Saudi Arabia

Dear Sir,

We desire to expand the amount of business we do with the Middle East, and would be most grateful if you could provide us with a list of reliable business concerns in your area which might be interested in importing Taiwanese Electronic Equipments.

We are well – organized exporters of Electronic and Hi – Fi products. Having been in business for more than 10 years, we are confident we can give your customers complete satisfaction.

As to our credit and financial standing, we can refer you to Bank of Taiwan, Head Office, Taipei.

Your assistance and early reply will be greatly appreciated.

                                                              Yours faithfully,

---

1. to expand：擴展。

2. The amount of business we do with the Middle East
   本公司在中東地區的業務量。

3. Would be most grateful if you could
   (煩請…爲荷)。類似的表達方式尚有
   shall / will be pleased if you will…
   shall / will be obliged if you will…
   shall / will appreciate it if you will…
   將 shall / will 改爲 should / would 則更爲客氣。

4. provide us with a list：提供名單。

請注意 provide A(人) with B(物)的句型。具類似意義及句型
的動詞尚有 supply, furnish 及 favor 等。

5. reliable business concerns：可靠的廠商。

　　business concerns：有關的廠商。

6. which might be interested in…：(可能)有意於。

7. electronic equipment：電子器材。

　　文意：本公司爲拓展在中東地區的業務，煩請貴會提供有意進口台灣
製電子器材的廠商名單爲幸。

　　第二段以組織的健全及令人滿意的服務，介紹本公司的優點。

1. well－organized：組織健全的，有基礎的。

　　利用 well 爲首的複合形容詞可以簡潔意思表達，如 well－advised
decision 明智的決定，well－balanced mind 精神正常，well－bal-
anced account 借貸平衡，well－informed man 消息靈通人士，
well－beloved 受人愛戴的，well－placed 地位適可，well－known
知名的。

2. exporters：出口廠商(係指經營的人員，故用複數)。

　　貿易英文常用的 importer(進口商)，customer(顧客)，buyer(買
方)，agent(代理商)，shipper(發貨人)，manufacturer(製造廠
商)，company(公司)，corporation(法人團體)，firm(廠商)，
bank(銀行)，factory(工廠)等集合名詞，其單數與複數之表示須
配合文意。例如：The company have decided not to send their
representative. 公司方面決定不派遣代表。因係指決策部門的人員
故爲複數。The company is financially strong。該公司財務健全。
因係指公司的結構，非人員，故而爲單數。

3. confident：有信心，有把握。

$$\left.\begin{array}{l} \text{We are confident of} \cdots \\[2em] \text{We are confident that} \cdots \end{array}\right\} = \left\{\begin{array}{l} \text{We feel confident of} \cdots \\ \qquad\qquad (對……有信心) \\ \text{We feel confident that} \cdots \end{array}\right.$$

文意：本公司乃組織健全的電子及身歷聲產品的出口商，經營已逾十年，服務週到，令人滿意。

第三段提供本公司的徵信銀行，以便對方查詢有關信用及財務狀況。

1. as to：關於，至於，用於句首有強調作用。

2. Credit and financial standing：信用及財務狀況。

3. refer you to(人)for(事)：有關…事宜請洽…。

文意：有關本公司之信用及財務狀況，請洽台灣銀行總行。

第四段以謹謝協助並早日賜覆結束。

貿易書信的結尾，尤其是帶有要求的書信，必須具有說服力，但不可失禮儀。本段相當於 We will appreciate your assistance and early reply. 但是第一段和第二段均以 We 開頭，故改用 you 口氣。

文意：祈盼貴會之協助，並早日賜覆爲禱。

㈡謝函(Thank you letter)

實踐貿易公司去函吉達商會之後，不久收到了該會提供的有意進口電子器材的廠商名單。

凡是委請的事項有了消息，禮貌上應回以謝函。謝函屬於被動回信的一種。回信的頭一段務必提及來信日期；一則便於文書上的處理，二則用來函日期代替來函的主旨，因此有利於書信的簡潔。茲舉數例如下以供參考：

We have just received your letter dated…

We are pleased to receive your letter dated…

We appreciate your letter dated…

Thank you for your letter dated…

以上各句均爲「　月　日大函敬悉」之意。

實踐貿易公司回覆吉達商會的謝函如下〔例5−2〕。

〔例5－2〕

Dear Sir,

We have just received your letter dated March 20, 1998 giving us the names of some possible importers of Electronic Products in your city.

Please accept our sincere thanks for your information, which will help us develop a new set of what should be mutually beneficial business relationship with some of firms in your area.

Sincerely yours,

第一段提及來函日期及其附件。

本段本來可以 Thank you for your letter 開頭，但是擬於第二段再致謝意，為免重複，乃改用現在完成式的 We have just received your letter 表示收到來信立即回覆。

1. giving us the names of…：惠賜…的名單。此乃來函的主旨，按一般的貿易書信，不必重述來函的主旨，此處係因謝函而刻意表達，令對方欣慰。

2. possible importers：未來的進口商。在尋找客戶的階段，任何對方都是有希望成為日後的往來對象，因此又稱 prospective importers。

文意：三月二十日大函惠賜電子器材進口廠商名單敬悉。

第二段致謝提供資料並展望未來。

1. please accept our sincere thanks for…：
   為…敬致謝意，謹謝…

2. help us develop… ＝help us (to) develop…：
   有助於本公司推展……

3. a new set of…business relationship：新的業務往來關係。同業間的

往來建立於買賣雙方，缺一不可，故而稱 set。

4. mutually beneficial business relationship：互惠的業務關係。

文意：為貴會所提供資料謹致萬分謝意。此項資料將有助於本公司與貴地廠商之間，發展新的互惠業務關係。

二、信用調查 (Credit Inquiry)

交易的對象，不論是自行接洽或經過介紹的，由於是未曾謀面，萬一遭遇惡劣的廠商或瀕臨倒閉的，或慣於詐欺的，不但無利可圖，而且得不償失。另一方面，對外貿易的各項費用必然較之國內交易為高，因此每一筆交易在交易量及金額方面均要求起碼的數量 (sizeable quantity)，否則划不來。但是隨着交易量的增加，商業上的風險亦隨之增大，故事先的信用調查是必要的。為防止或減輕此項風險，調查的方法是：向同業間打聽對方的信用狀況，或洽詢往來的銀行，前者為備咨商號 (Trade Reference)，後者為備咨銀行 (Bank Reference)。我國的廠商在國內，可以透過有往來的外匯銀行，或徵信機構，或逕洽對方指名的外國銀行，取得有關的徵信資料。

· 函請外國銀行徵信調查

實踐貿易公司接獲香港進口商的來信，欲購貨品並指名美國銀行香港分行為備咨銀行。實踐貿易公司於是去函委託徵信如下〔例 5-3〕。

本函屬於主動的委託函，應提介紹人或關係人作為開頭。

第一段略提被調查的公司與銀行及調查人之間的關係。

1. a request for supplies of：…要求供應…，欲採購…

2. on D/A terms ＝on Documents against Acceptance terms：以承兌交單方式付款，乃託收方式之一，進口商憑承兌匯票即可提領貨品，而於匯票到期日再行付款。託收方式之交易對象更須信用調查。

3. given us your name as a reference：指名貴行為備咨人。

〔例 5－3〕

The Manager
Bank of America
Kowloon
Hongkong

Dear Sir,

Audio Importers Co., Ltd., Hongkong have sent us a request for supplies of our products on D/A terms, and given us your name as a reference in regard to their financial stability.

We should be greatly obliged if you would favor us with your confidential opinion as to their financial standing and the scope of their transactions.

Any information you may give us will be treated as strictly confidential. We thank you for your assistance.

Yours very truly,

4. in regard to ＝concerning：關於。

5. financial stability：財務結構。

一般的徵信內容有(1)信用狀況（Credit standing or status）；(2)營業狀況（Business standing）或營業績效（business turnover）及(3)業務經驗（Business experience）。

文意：香港的奧迪公司欲以承兌交單方式進口本公司產品，並指名貴行為該公司財務狀況的備咨銀行。

第二段請求銀行提供徵信資料：

1. We should be greatly obliged if you would…：煩請。（參閱〔例 5－1〕）

2. favor us with…：惠予。（參閱〔例 5－1〕）

3. confidential opinion：銀行在徵信資料中所提供的意見僅對當事人

而不對外公開，故爲機密性的意見。

4. as to：至於。（參閱〔例5－1〕）

5. financial standing：財務狀況。

6. the scope of their transactions：該公司的業務展望。

文意：煩請貴行就有關該公司的財務狀況及業務展望方面惠示卓見。

第三段對徵信內容具結保密。

委託徵信調查的函件，依例應由委託人對備咨人具結(1)保密及(2)免責兩項條款。

1. will be treated as trictly confidential：視爲機密；以密件處理。本函遺漏了免責條款，應加…and without any responsibility on your part(並與貴行無涉)，始爲一完整的具結文。

2. We thank you for your assistance：對於貴行的協助謹致謝意。爲配合 you 口氣，本句改爲 Your assistance will be appreciated 則更佳。

文意：貴行所提供的資料當視爲密件並謹謝貴行的協助。

對於廠商的委託查詢，銀行通常覆以徵信資料(Credit Information)或徵信報告(Credit Report)。其內容不外被調查公司行號的創業年代(Year of Establishment)、資本額(Capital)、負責人(Responsible Person)、營業項目(Line of Business)、業務量(Sales)、信用狀況(Credit Standing)並附簡短的客觀評語，提供查詢者研判對方信用狀況的參考。

銀行在覆函中有提供具體資料的義務，但有不對委託人負任何責任的權利，因此均以免責條款結尾。例如：" This information is given in strict confidence and without responsibility on the part of this bank or any of its officers. "或" This information is PRIVATE AND CONFIDENTIAL and furnished as a matter of business courtesy in reply to your inquiry. No responsibility is assumed by this bank or any of its officers. "

三、招攬業務(Trade Proposal)

經過徵信調查之後，對於信用狀況良好的廠商，我方即可提議進行交易，俗稱招攬業務。招攬業務的書信猶如相親和面談，必須令對方產生良好的印象，以便爭取業務。其內容依序為

(1)　認識的經過

(2)　自我介紹

(3)　提議交易的項目

(4)　交易有關的條件

(5)　信用備咨人及

(6)　結尾。

澳洲有一家照相器材進口商，於參觀最近在西德科倫舉辦的照相器材樣品展覽會後，中意其中的三角架，於是致函展示該三角架的廠商，表達有意與之交易如〔例5—4〕。

本函雖然未提及介紹人或關係人，但是在開頭述及受信人在樣品展覽會場展示的產品。這不但能令受信人為展示有了效果而感到欣慰，更能促進彼此的親切感，是很有力量和味道的開頭。

第一段簡介自己並提及交易的標的物。

1. As…, we are very interested in…：由於…，本公司對…非常有興趣。to be interested in 在此相當於「有意購買」。

2. leading importers：主要的進口商。

3. wholesalers：批發商，零售商為 retailers。

4. photographic equipment：照相器材，equipment 為集合名詞，限用單數。

5. professional type tripods you displayed：貴公司所展示的專業用三角架。

6. Photo—Kina in Cologne：在西德科倫舉行的照相器材樣品展。

〔例 5－4〕

Gentlemen:

As one of the leading importers and wholesalers of photographic equipment in Australia, we are very interested in the professional type tripods you displayed at the recent Photo – Kina in Cologne.

There is a steady demand here for these types of tripods, especially the high quality ranges and fashionable desings. We have been receiving a number of enquiries from our trade connections in this area for these tripods and think we may be able to place regular orders with you if your prices are competitive.

Will you please, therefore, quote us your lowest prices CIF Sydney and let us know your delivery schedule. For your information, we have enclosed our latest " Introduction to Prophonic ".

We look forward very much to hearing from you.

Yours faithfully,

文意：由於本公司係澳洲的進口兼批發照相器材的主要廠商之一，有意訂購貴公司在科倫的照相器材樣品展覽會場上所展示的專業用三角架。

第二段為爭取有利的業務，提供澳洲當地對此項器材的需要市場的現況，並提議訂貨數量。

1. a steady demand：穩定的需要量，形容市場的需要量保持平穩，言外之意即銷路可觀。

2. high quality ranges：高級系列產品。

3. fashionable designs：時髦的式樣。

4. we have been receiving a number of enquiries：本公司陸續接獲大批詢價信。

   have been receiving：用現在完成式，充分表達繼續不斷的狀況。

第五章 交易的準備

enquiries =letters of enquiry：詢價信，澳洲是英語系統的國家，慣用 enquiry，而美語系統的國家地區，則使用 inquiry。

5. trade connections：交易的主顧，即往來客戶。此際係指與該公司有業務往來的零售商。

6. regular orders：定期性的訂單，即每半年或每三個月或每個月固定發訂單。訂貨的動詞爲 to place，因此向貴公司訂貨爲 to place an order with you。

7. prices are competitive：價格公道、合理，原意爲能與其他同業競爭得過的價格。

文意：此地對於上述的三角架，尤其是高級品系列和時髦式樣的三角架之需要量甚爲平穩。本公司陸續接獲很多往來客戶，有關此類三角架的詢價函，因此認爲只要貴公司所定價格合理，當可定期向貴公司訂貨。

第三段函請報價並惠示交貨時間，附贈雜誌供閱，祈盼回音。

1. Will you please…：煩請…，敬請…。在貿易英文的書函裏使用 will you please 時，其句尾並不加疑問號「？」。

2. quote：估價，名詞爲 quotation（估價單）。

3. lowest prices CIF Sydney：最低的雪梨港起岸價格，即到雪梨港包含運費、保險費在內的最低價格。lowest prices 最低價格又可稱 rock —bottom prices 或 best prices。

4. delivery schedule：交貨時間表，指預定交貨的時程。

5. For your information：參考起見，謹供參考。

6. we have enclosed…：隨函檢送。相同的表達方式尚有

(1) We enclose our annual report.

(2) We are enclosing our Quotation No. 71 −02.

(3) Enclosed is our illustrated catalog.

(4) Enclosed please find our price list.

以上四句中，第三句是比較簡要的表達方式，係利用被動語態的倒置法；原句型爲 Our illustrated catalog is enclosed，第四句是比

較古老的表達方式，乃命令句 Find our price list enclosed 的倒置，不宜常用。

7. "Introduction to Prophonic"：業務簡介的書刊名。凡引用書籍、刊物、雜誌等名稱時常用引號（"　　"）或劃線(—)表示。

8. We look forward to…：盼望…，祈盼…，是幸。

9. hear from you：收到回信，惠覆。

文意：為此請貴公司惠予估價，最低的雪梨港起岸價格，並預計交貨時程。隨函檢送"Introduction to Prophonic"乙冊，敬請參照見復為幸。

四、回覆招攬業務函(Response to Trade Proposal)

對於國外廠商來函招攬業務，受信人應充分了解來函的主旨，考慮對方市場的動態，並配合本公司的立場，適時地予以回覆。有意與之交易時，不可操之過急，仍須事先調查對方的信用狀況，以免落入專門假藉招攬生意索取樣品者(Sample Collector)的圈套，而枉費一筆可觀的樣品以及包裹郵費。來意不清楚之處，務必去信澄清，不可自作主張，以免日後引起糾紛。

由於招攬業務的書信，亦屬於詢價函的一種，因此肯定的答覆相當於報價，故而必須具備各項國際貿易的基本交易條件，詳情參閱第六章。在此為讀者提供否定的答覆例〔例5−5〕。前述在科倫照相器材樣品展覽會參展的製造廠商，雖然參展的目的在於廣招生意，但是該廠商的銷售方式是採取總代理制，因此無法答應做直接的交易。該製造廠商在回信中雖然婉謝直接交易，但是毫無否定消極的語氣，反而是一封充滿溫情的書信，甚多可取之處。

第一段對於來函查詢表示謝意。貿易書信不宜在頭一段即表明謝絕或唱反調，除非不得已須在頭一段表示回絕，亦不可擺在第一句，應挪到第一句之後或第二句。本函將婉謝的意思移到第二段表達。

1. inquire about：查詢，即打聽產品。

〔例 5－5〕

Dear Sirs,

We wish to thank you for your letter dated June 10, 1998 inquiring about our SHARP brand tripods.

Although we highly appreciate your interest in our products, we are sorry that we are unable to meet your request for a direct offer. Our business in Australia is now handled on an exclusive basis by the following firm:

> Australia Prophoto Pty. Ltd.
> Challis House
> 4－10 Martin Place
> Sydney, NSW 2000
> Australia

As we believe your inquiry will definitely interest the above sales company, we have duly passed your letter onto them with instructions for them to immediately contact you.

We hope you can work out a successful business relationship with Australia Prophoto Pty. Ltd. Thank you again for your interest in SHARP Photographic Equipment.

> Yours very truly,

文意：貴公司六月十日大函查詢本公司 SHARP 牌三角架乙節謹誌謝意。

第二段將無法直接回覆有關查詢內容的具體理由提出，以便對方諒解。

1. highly appreciate your interest：銘謝貴公司的來意。

2. we are unable to meet your request：歉難答應。

3. direct offer：直接報價，即不透過代理商，由賣方直接向買方報價者。

4. on an exclusive basis：總代理制。由出口廠商委託某特定地域內的一家廠商獨家銷售，此際受託人為總代理商 (Sole selling agent)。總代理契約一經成立，委託的出口廠商在該地區出售貨品，必須全部經過該總代理商報價，不得直接或經過第三者向該地區購貨者報價。

5. Pty. Ltd. (Proprietary Limited)：英國公司法上的股份公司，常見於新加坡、馬來西亞、澳洲、紐西蘭等地的公司行號。尚有 Pte Ltd. (Private Limited)亦屬於公司組織的一種。

文意：對貴公司洽詢本公司的產品乙節甚為銘謝，惟歉難直接報價。本公司在澳洲的業務目前係委託下列廠商為總代理：

澳洲專業照相器材公司

澳洲，新南威爾斯州

雪梨市，馬丁廣場

查理大廈

第三段：雖然無法交易，但是對於業務的爭取仍然有利，於是將來函轉遞總代理商，並囑其儘速連繫。

1. pass…onto…：將…轉交給…。

2. with instructions：附帶要求(指示)。

3. contact：連繫，例如 I thought I had better contact you at once. 惟此際連繫的方式未交代清楚；若用電話連絡則寫成 I thought I had better call you by telephone. ；用書信連絡則為 I thought I had better write you.。

文意：鑒於貴公司的查詢，足以令上述經銷商感到興趣，本公司已將來函轉遞該公司，並囑咐立即與貴公司接洽。

第四段：如此一來經銷商和該公司之間可望達成交易，則間接地促進產品的銷售，故希望他們能建立往來關係並再致謝。

1. Work out：做成，完成

文意：祈盼貴公司和澳洲專業照相器材公司之間建立成功的業務關

係，謝謝貴公司查詢 SHARP 牌產品。

練習問題 ●————————————————————

一、試述從事外銷，如何招攬業務。

二、本公司的產品在品質上有些差異時，應寄送何種樣品爲宜，試述理由。

三、當出口部門的業務員應有的態度如何，試述你的見解。

四、試指出下列各句的主旨，敘述重點即可。

1. We have been engaged in overseas trade since 1947.

2. It is likely that they will be in the market shortly.

3. We assure you that they sell quite well in your market.

4. We are arranging with the manufacturers to meet your requirements.

5. We understand that your government contemplates import of this material soon.

6. Our above prices may be improved upon receipt of a specific inquiry (an inquiry with details).

7. Demands during the last two months have been quiet due to overstock among our competitors.

8. We have been dealing in (carrying on) this line of business for the past 10 years with fair records and reputation.

五、詳閱下列書函並回答各項問題：

Dear Mr. Sullivan:

We are glad to know of your interest in the Eastman Kodak Company and would like to send you a copy of our annual report. We are also forwarding four issues of " Highlights, " our quarterly publication to shareholders, which will bring you up — to —date on Kodak operations.

In order to give you further background on the company, we are enclosing a booklet which describes one of our manufacturing plants, a copy of our latest sales and earnings release, a brief company history, and several other brochures.

We are pleased to have this opportunity of introducing you to Kodak, and if we can be of any assistance in the future, be sure to let us know.

Sincerely,

1. 試指出各段的主旨。

2. 本函是否屬於回信？寫給誰的？是誰寫的？

3. 本函有無附件？那幾件？

4. 本函的用意何在？

5. 假使你收到此信，你如何回覆？簡述你的處理方式？

六、根據下列情況草擬一封委託介紹客戶的信。

1. 本公司擬進口貴國的辦公用事務機器(Business Machines for Office Use)。

2. 請貴商會(Chamber of Commerce)推介殷實的出口商。

七、根據下列內容準備一份委託徵信調查的信。

1. 本公司將與美國市場公司(American Market Corp., New York)建立業務關係。

2. 該公司提貴行為備咨銀行。

3. 請提供該公司的徵信資料。

4. 對貴行所提供的資料，本公司具結保密並與貴行無涉。

Trade English 第六章

交易的進行

一、詢價──有關的交易條件

進口商與出口商之間經過初步的接觸，認識彼此的交易原則之後，即可進行個別項目的買賣。進口商方面雖然由索取的樣品或目錄以及價目表可得到交易項目的概念，但是具體的細節尚需澄清和確認。因此在這一階段，進口商往往向出口商查詢有關貨品的品質、價格以及其他各項條件。此類函件的英文稱為 Inquiry 或 Enquiry，而在我國則譯成詢價，其實 Inquiry 可查詢的內容有八項：

- (1) 貨品 (Commodities)
- (2) 品質 (Quality)
- (3) 數量 (Quantity)
- (4) 價格 (Price)
- (5) 交貨 (Shipment)
- (6) 包裝 (Packing)
- (7) 付款 (Payment)
- (8) 保險 (Insurance)

惟查詢者和被查詢者，均須對查詢的內容應有基本上的認識，方有助於草擬詢價函或答覆查詢，以下先簡述各項查詢內容的概念供讀者參考。

(一)貨品 (Commodities)

1.貨品名稱 (Description of Commodities)

交易的標的物通稱為貨品，而每一項貨品均有其固有的名稱和統稱。固有名稱，除了少數貨品係由進口商方面所設計而交由出口廠商製造者之外，通常以出口商所定或提出的名稱為準。至於統稱則以 Commodities 和 Merchandise 兩字為正式的統稱，前者有總稱的涵義，故使用範圍最廣；後者則因具有商品的意義而常用於經銷商或貿易商之間。其他尚有 Goods

指某類貨品，如 leather goods 指皮貨，Product 指產品；Material 指材料和原料；Supplies 指供應品；Article 指成衣類或家電用品等貨品；Item 則指單項的貨品。不論任何貨品，就船公司的立場而言均視為 Cargo（貨物）。

2.貨品分類（Classification of Commodities）

我國的進口商在查詢某項貨品之前應認識該項貨品的類別和分類標準。根據國際貿易局的規定，我國進出口貨品分為(1)准許類——包括准許進口類和准許出口類；(2)管制類——包括管制進口類和管制出口類；(3)禁止類——包括禁止進口類和禁止出口類。將來申請進出口時應依照國貿局公布的貨品類別向國貿局或其委託之簽證機構申請簽發輸入或輸出許可證。此際所填類別稱為中華民國進出口貨品分類（Classification of Import & Export Commodities of the Republic of China）又稱國家分類標準（CCC code: China Commodity Classification Code）。此外就運輸方面而言，貨品又分一般貨物（General Cargo）和特殊貨物（Special Cargo）。前者為無須使用特別方法裝卸積載或處理的貨品總稱，因有雜類貨品之意義，故俗稱雜貨。後者為須用特殊的設備工具以之裝卸積載並加以特別照料的貨品。

(二)品質（Quality）

品質是交易標的物的貨色。貨色的標準和決定為交易成交的關鍵要素之一。品質的標準依貨品的交易習慣而定，例如以貨樣為標準者，此項交易方式屬於 Sale by Sample（憑貨樣買賣）。貨樣又分 Seller's Sample（賣方貨樣）、Buyer's Sample（買方貨樣）及 Counter Sample（相對貨樣）。屬於憑貨樣買賣的尚有 Sale by Pattern（憑花樣買賣）、Sale by Design（憑圖樣、設計買賣）、Sale by Model（憑模型買賣）及 Sale by type（憑類型買賣）等，應用範圍最廣。得與其他方式併用的則有 Sale by Description（憑說明買賣），例如機械、音響器材等以圖樣配以照片，加上規格（Specification）說明。某項貨品因其知名度已高，因而指定廠牌作為貨樣的標準者乃 Sale by Brand or Trade Mark (憑商標、品牌買賣)；至於無法一一說明或大宗物資，則採 Sale by Standard or Grade (憑標準物買賣)。以上各種交易方式下的貨樣標準於何時決定，亦為進出口廠商雙方所不可忽略者，通常採取 Shipped Quality Final (裝運品質為準)，遇有特殊貨品時始用 Landed Quality Final (卸貨品質為準)。

(三)價格 (Price)

貨色好固然令人動心，價格公道低廉更能具體地把握購買者。因此價格往往構成交易成交的決定性關鍵。對外貿易的價格以單位價格表示，例如：US$25.50 per dozen CIF New York，其涵義包括交易使用的幣別 US$，單位價格的金額 25.50，交易的單位 per dozen 以及貿易條件 CIF New York 等，缺一項即不為完整的價格表達。

1.貨幣 (Currency)

(1)外幣 (Foreign Curreney)

對外貿易收付貨款 (Proceeds)，涉及外匯。我國的外匯管理採申報制。新台幣 50 萬元以上之等值外匯收支或交易，應依規定申報。出口所得之外匯應結售中央銀行或其指定銀行，或存入指定銀行，並得透過該行在外匯市場出售。至於進口所需支付之外匯，得自存入外匯自行提用，或透過指定銀行在外匯市場購入，或向中央銀行或其指定銀行結購(以上參閱附錄「管理外匯條例」第 6 條、第 7 條及第 13 條)。目前各指定銀行掛牌買賣的幣別常見的有：美金(U. S. Dollar)、澳幣(Australian Dollar)、奧地利幣 (Austrian Schilling)、比利時法郎 (Belgian Franc)、加拿大幣 (Canadian Dollar)、馬克 (Deutsche Mark)、法國法郎 (French Franc)、港幣(H. K. Dollar)、日圓(Japanese Yen)、馬來西亞幣(Malaysia Ringgit)、荷蘭幣 (Netherlands Guilder)、紐西蘭幣 (New Zealand Dollar)、英磅 (Pound Sterling)、新加坡幣 (Singapore Dollar)、南非幣 (South Africa Rand)、瑞典幣 (Swedish Krona)、瑞士法郎 (Swiss Franc)、泰銖 (Thailand

Baht)及歐洲通貨單位(ECU)等。另外自 1999 年 1 月 1 日起新的貨幣歐元 (EURO)開始流通(詳附錄六)。

(2)匯率(Exchange Rate)

上述各種外幣,由各指定銀行(俗稱外匯銀行)於營業日上午九時起掛牌,按買入匯率(Buying Rate)和賣出匯率(Selling Rate)買賣。外匯的匯率又分即期匯率(Spot Rate)與遠期匯率(Forward Rate)。即期匯率即買賣當日交割的匯率;遠期匯率之期別通常分 30 天、60 天、90 天、120 天及 180 天期,買賣的幣別及匯率由外匯銀行視市場行情掛牌或與客戶逕行議價。

2.單位(Unit)

交易的單位以數量(Quantity)表示。數量分重量(Weight)、長度(Length)、個數(Number)、容器或包裝(Container or Packing)、面積(Area)、容積(Capacity)及體積(Volume)等。使用重量單位的貨品最多,重量分公制(Metric System)和英制(British System)。一般貨品使用的英制重量為常衡(Avoirdupois),而金、銀、寶石等則使用金衡(Troy)。長度亦有公制與英制之分。個數的單位中常用的名稱有 Piece(件,個,張)、Set(套,組,列)、dozen(打)、ream(令)及 coil(線圈)等。容器或包裝為單位者如水泥、麵粉等以一包(sack)一袋(bag)計算買賣。面積單位使用於木板、皮革等。容積則用於穀類、液體物質等,而體積則用於氣體、瓦斯等。

使用數量單位應注意:(1)同一貨品使用的數量單位未必相同,例如小麥的買賣依地區習慣的不同而分別使用 Bushel(蒲式耳)、磅(Pound)、噸(ton)及公斤(kg)等單位。(2)單位名稱相同者,其值未必相等。例如:「噸」的單位有公噸(Metric Ton)為 1,000 公斤,英國的長噸(Long Ton)為 1,016.064 公斤,而美國的短噸(Short Ton)為 907.18 公斤。上述的蒲式耳單位在美國秤玉米、小麥和黃豆及其他穀類時又分別使用 56 磅、60 磅及 62 磅等不同重量。(3)計數單位和計價單位未必相同,例如夾板的計數單位爲 Sq. ft(平方英呎），但是計價時卻以 100 平方英呎或 1,000 平方英呎爲單位。

以重量爲計價單位的貨品，其重量的決定地點與時間通常採用裝貨地裝貨時的重量爲準(Shipped Weight Final)，亦有少數貨品採取卸貨重量爲準(Landed Weight Final)者。計量的方法通常分淨重(Net Weight)和毛重(Gross Weight)，後者係淨重加上皮重(Tare)的總和。偶有要求以純淨重(Net Net Weight)表示者，此乃淨重減去內包裝後的重量。

3.貿易條件(Trade Terms)

貿易條件爲貿易術語，使用特定的用語或縮寫字表示下列各項內容：(1)進出口貨品價格的構成；(2)買賣雙方應履行的義務；(3)貨品的所有權及風險轉移的時間與地點；以及(4)各項費用的負擔界限等。依國際商會於 1990 年修訂的國貿條規(INCOTERMS =International Commecial Terms，其正式名稱爲「貿易條件的國際解釋規則 International Rules for the Interpretation of Trade Terms)，有以下 13 種貿易條件。

(1) EXW =Ex Works： 工廠交貨條件

賣方負責把貨物在其原地，如工廠、廠房或倉庫等，交給買方。從此地起運到目的地爲止之所有費用和風險由買方負擔。

(2) FCA =Free Carrier：指定地點交貨條件

賣方應完成運交貨物以爲出口，而於買方所指定地點或場所交貨。若買方未指明正確之交貨地點，則賣方得選擇在運送人收取貨物的地點接管。本條件也適用於任何運輸方式，包括複合運輸方式。

(3) FAS =Free Alongside Ship：船邊交貨條件

賣方負責將貨物運至輸出口岸所指定的碼頭或駁船之船邊，其後之所有費用和風險均由買方負擔。

(4) FOB =Free On Board：船上交貨條件

賣方負責將貨物交付至裝貨港之指定船隻，並越過船舷欄杆爲止。買方則承擔所有成本和自起運點所發生之風險。

(5) CFR ＝Cost and Freight：運費在內交貨條件

　　賣方負責支付運送貨物到指定之目的地港口的成本和費用。貨物越過船舷欄杆時起所產生之風險及任何增加之費用由買方負擔。

(6) CIF ＝Cost, Insurance and Freight：運保費在內交貨條件

　　賣方依約投保海上保險並支付保險費之外，應履行與 CFR 相同之義務。

　　對我國的進出口廠商而言，以上之 FAS、FOB、CFR 及 CIF 四種貿易條件僅適用於海運。

(7) CPT ＝Carriage Paid To：運費付訖交貨條件

　　賣方負責辦理貨物之出關手續，並支付貨物運至指定目的地之運費。貨物交給運送人經營之後所發生之任何額外費用及風險歸買方承擔。本條件適用於複合運輸在內之任何運輸方式。

(8) CIP ＝Carriage and Insurance Paid to：運保費付訖交貨條件

　　賣方除負責 CPT 條件下相同義務之外，並須為買方辦理投保手續且支付保險費用。本條件適用於包括複合運輸在內的任何運輸方式。

　　以上 8 種貿易條件因賣方的履行義務均在出口地，故屬於輸出地交貨貿易條件。至於下列 5 種貿易條件則為輸入地貿易條件，實務上不適合於海島經濟型的我國廠商。

(9) DAF ＝Delivered at Frontier：邊境交貨條件

　　本條件主要使用於貨物以鐵路或陸路方式運輸。賣方備妥貨物並負責辦理通關運交至邊境所指定之地點，但在鄰國的海關之前。

(10) DES ＝Delivered Ex Ship：到岸船上交貨條件

　　賣方必須負責將貨物運至指定之目的港，在辦理進口通關之前的船上交給買方。至指定之目的港為止之費用及風險均歸賣方。本條件適用於海運。

(11) DEQ ＝Delivered Ex Quay：碼頭交貨條件

　　賣方負責將貨物運至買方指定之目的地的碼頭，並辦妥進口通關手續，賣方負擔所有費用及風險。若賣方無法直接或間接辦理通關手續，則

不適用本條件。

(12) DDU =Delivered Duty Unpaid：稅前交貨條件

賣方負責將貨物運至輸入國境指定之地點，並辦理通關手續。有關之運費及風險歸賣方負擔，惟不包括進口關稅、稅捐及其他規費。

(13) DDP =Delivered Duty Paid：稅後交貨條件

賣方的責任除與 DDU 的部分相同之外，所負擔的費用包括關稅、稅捐及其他通關所需支付之規費。

(四)付款 (Payment)

成交貨品的貨款 (Proceeds) 將來如何給付，也是買賣雙方關心的事宜。對外貿易支付貨款分匯付和出票兩種方式。匯付 (Remittance) 屬於順匯，其方法有票匯 (D / D: Demand Draft)、信匯 (M / T: Mail Transfer) 和電匯 (T / T: Telegraphic Transfer)。我國的出口廠商和貿易商接獲小額訂單時，常遇以上的方法收受貨款。出票 (Drawing) 則屬於逆匯，乃利用匯票 (Draft) 的功能以收回貨款的方法。匯票憑票本身向被出票人要求償付款項者為光票 (Clean Draft)；而需檢附有關單據始能要求償付貨款者為跟單匯票 (Documentary Draft)。後者又分為憑信用狀出票 (Drawn under L / C) 和託收匯票 (Bill for Collection)。信用狀項下的匯票依付款期限而分即期 (Sight) 和遠期 (Usance)。託收匯票則有付款交單 (D / P: Documents against Payment) 和承兌交單 (D / A: Documents against Acceptance) 之別。

(五)交貨 (Shipment)

有些貨品為配合季節性或訂戶的需要，須預估裝運日期 (Shipment Date)。此際應確切表示預估的裝運日期如：July Shipment, Shipment During July 或 On or Before July 30, 1998，但不宜使用立即 (Immediate 或 Prompt) 或儘速 (As Soon As Possible) 等不具體的用語。運輸工具則除海運 (by Vessel) 之外尚有空運 (by Airlift) 和包裹 (by Parcel Post)。海運方面，一般雜貨類均委託定期船 (Liner) 交貨，又分一般海運和貨櫃運輸

（Container Shipment），至於大宗貨品（Bulk Cargo）則交不定期船（Tramp）運輸。

㈥包裝和刷嘜（Packing & Marking）

對外貿易對一般雜貨的運輸交貨要求適當的包裝，即該貨品習慣上耐航的包裝，此際所稱的包裝係指外包裝（Packing）而非內包裝（Packaging），乃爲保護貨品於運輸和保管上所作的包裝。小麥、黃豆、玉米等穀類和礦砂等不另包裝或不必包裝者屬於散裝貨（Bulk Cargo），鋼鐵、錫塊等形態上自成件數者，因毋需包裝而被稱爲裸裝貨（Nude Cargo）；至於一般雜貨均需包裝者，屬於包裝貨（Packed Cargo）。包裝貨依其包裝材料和形態上的不同，而大別爲：(1)箱裝（Case Packing）；(2)梱包（Bale Packing）；(3)袋裝（Bag Packing）；(4)桶裝（Barrel Packing）；(5)瓶裝（Bottle Packing）；(6)簍裝（Basket Packing）；(7)籠裝（Cage Packing）；(8)捲裝（Coiled Packing）以及(9)貨櫃裝（Container Packing）等。

貨品經包裝後，需在包裝容器上加印識別船貨的嘜頭或標誌（Mark or Shipping Mark）。嘜頭分主標誌（Main Mark）、目的港標誌（Port or Destination Mark）和箱號（Case Number）。最常見的主標誌爲三角形（△Triangle）和菱形（◇Diamond）。

SHIPPING MARKS：

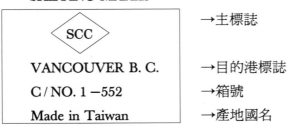

→主標誌

→目的港標誌
→箱號
→產地國名

㈦保險（Insurance）

貨品在運輸途中可能遭遇的風險，得因保險而獲得補償。常見的保險種類有：(1)平安險（FPA: Free from Particular Average），即單獨海損不賠；(2)水漬險（WA: With Average）即單獨海損賠償險；(3)全險（AAR: A-gainst All Risks 或 AR: All Risks），全險雖然對於運輸途中，因外來因素所致的一切損失負責賠償，但是排除因滯延、固有瑕疵及戰爭、罷工、暴亂等所引起的損害賠償。因此而產生的附加保險條款有：(1) War Risks（兵險）；(2) SRCC: Strikes, Riots and Civil Commotions Clauses（罷工暴動險）及(3) TPND: Theft, Pilferage and Non－Delivery（偷竊及未送達險）。

上述之全險、水漬險和平安險，自倫敦保險市場於一九八二年起改用表列式的新海上保險單之後，分別由協會貨物條款(A)、協會貨物條款(B)及協會貨物條款(C)所取代。其簡稱爲 ICC (A) [Institute Cargo Clauses (A)]、ICC (B)及 ICC (C)。

二、詢價函（Inquiry Letter）

美國的腳踏車經銷商，有意進口在雜誌上廣告的台灣製腳踏車，因而來函查詢有關的交易條件〔例 6－1〕。

第一段提及廣告並有意進口男女用腳踏車。

1. We：指本公司。貿易書信中，公司行號機構等均以複數表達，因此 you 指貴公司，而 they 則指該公司。

2. January issue：元月號，issue 爲定期刊物雜誌的發行月份或號數。

3. "World Cycling"：「腳踏車世界」，刊物或雜誌的名稱通常以大寫開頭，並加引號（" "）或劃線（—）。

4. Taiwanese make：台灣製(名詞)；Taiwanese made 台灣製的(形容詞)，等於 of Taiwan made; made in Taiwan 指原產地爲台灣，即台灣製。

文意：本公司因翻閱貴公司刊載於元月號「腳踏車世界」的廣告，而對貴公司台灣製男、女及小孩用腳踏車頗感興趣。

第二段簡介公司的經營情形。

〔例 6－1〕

Gentlemen:

We have seen your advertisement in January issue of " World Cycling " and are interested in your bicycles of Taiwanese make for both men and women, and also for children.

As we are one of the leading bicycle dealers and have many branches in the States, we are in a position to handle large quantities.

Please quote us for the supply of the items listed on the estimate sheet enclosed, giving your lowest possible C. I. F. San Francisco prices. Will you please also state your earliest shipping time, your terms of payment, and discounts for regular purchases.

If the quality of your machines is satisfactory and the prices are right, we expect to place regular orders for fairly large numbers.

We look forward to your early and favorable reply.

Sincerely yours,

---

1. leading ＝first class ＝A1：一流的；主要的。

2. the States：係指美國本土。

3. to be in a position to：得…；能…，意義相當於 be able to，但語氣比較柔和。

4. to handle：經銷，經手。

文意：本公司具有大量銷售腳踏車的能力，蓋本公司爲美國主要的腳踏車經銷商之一，並且分公司遍及全美之故。

第三段檢附估價單請求報價。

1. to quote：估價，定價，亦作開價。quoted value ＝current price 爲時價；quotation 則指行市和估價單或報價單。

2. item：品名；貨品；細目。商品的稱呼尚有 commodity, merchan-

dise, goods, product 及 article 等，參閱(本章一，㈠貨品)。

3. estimate sheet：估價單。

4. lowest price: 最低價格。指價格不貴，便宜時宜用 competitive price, reasonable price 或 rockbottom price。

5. CIF San Francisco：到舊金山運費、保險費在內貿易條件。

6. Shipping time：裝運日期。對外貿易的交貨時間，在實務上均以賣方履行交貨義務的時間爲準。故通常所謂交貨時間即指裝運時間 (Time of Shipping)。

7. terms of payment：付款條件。Terms 一字可解釋爲條件、付款條件和術語；Terms & Conditions 則爲交易條件，其中的 Conditions 指付款以外的各項條件。

8. discount：折扣，減價，貼現。在美國 discount house 指廉價商店。

9. purchase：購買，進貨。make a purchase 採購。

文意：煩請就所附估價單上所列品名以到舊金山運費、保險費在內貿易條件報最低價格，並列明最早的裝運日期、付款條件及定期採購可享有的折扣爲幸。

第四段提醒品質和價格合意時將大量訂購。

1. machine：由電力或人力推動的機器，此際乃指腳踏車。

2. to be satisfactory＝to be sufficient：令人滿意；沒話說。

3. prices are right：價格適可，即買賣雙方均合意的價格。

4. to place an order：訂貨。

5. fairly：相當地。a fairly large car＝not small but not very large car 相當大的汽車。

文意：若貴公司的腳踏車品質夠格而價格合適，則本公司大量訂購可期。

第五段盼早日回音。

1. favorable reply：好消息。

文意：敬請早日惠賜佳音為荷。

三、回覆詢價函（Response to Inquiry）

　　回信的要訣很簡單，對方問甚麼答甚麼，對方的要求能否答應均應具體地答覆。前函〔例 6－1〕美國的經銷商要求估價並請示裝運日期、付款條件以及折扣。由來函第二段得知該經銷商分號遍及全美，大量訂購可期，則以 CIF San Francisco 估價應不成問題，裝運日期可洽船公司安排，付款條件當然要求憑信用狀，至於折扣在薄利多銷的原則下似可給予優惠。現在讓我們看看這家腳踏車廠商如何覆文〔例 6－2〕。

〔例 6－2〕

Gentlemen:

We thank you for your inquiry of February 10, 1998 and are glad to hear that you have seen our advertisement in January issue of World Cycling. No doubt you also read the item about our products on page 18 of this Journal. As you see our products are enjoying a good reputation all over the world. We pride ourselves on superior quality of our products and would be confident that they will help you expand your market.

We are separately sending you a copy of our latest price list giving CIF San Francisco prices together with an illustrated catalog for your reference. And shipment of the products could be made immediately upon receipt of a firm order.

Although the prices quoted are the lowest and leave us little margin of profit, we allow special discount of 3% on the order of one thousand bicycle and over in order to secure your initial order.

We look forward to the opportunity of being of service to you soon.

Very truly yours.

第一段致謝來函並強調產品優良。

1. We thank you for… : 謹謝…。表達感謝的語句尚有：

   We are grateful for… ; We appreciate…

   We are obliged to you for… ; We received your … with many thanks。

2. no doubt : 無疑地。

3. product : 產品。produce 為農產品。

4. journal : 定期刊物。此際係指 World Cycling。

5. to enjoy a good reputation : 獲得好評；to enjoy a good sale 暢銷。

6. all over the world : 全世界。

7. to pride oneself on… : 以…自豪，自傲。

8. superior quality : 最高品質。

   best, highest, superior, first class, A1 等用於形容最高，特等，最上，特級，優秀等品質。

   extra, excellent, prime 形容優良，上等，一級等品質。

   good, standard, medium, choice, average, fair 等形容好、標準、中等、特選、平均、相當好的品質。

   poor, cheap, inferior, bad 則形容劣等，廉價，低級，不良的品質。

9. to be confident : 有自信；有把握。

10. to expand one's market : 擴張市場；擴大銷路。

文意：二月十日大函敬悉，本公司於元月號「腳踏車世界」上的廣告，承蒙貴公司參閱，甚感榮幸。本公司產品在世界各地均博得好評乙節，諒閱讀該號第十八頁刊登的產品欄即可知。本公司自豪於產品品質的優秀並確信可助貴公司推展銷路。

第二段通知另郵寄價目表檢附目錄。

1. to be sending : 寄送。在貿易英文常以進行式強調動態。

2. separately 形容動詞 send 而表示「另郵」，相當於 under separate

cover; by separate mail。

3. latest：最新的；最近的。

4. illustrated catalog：附圖解和說明的目錄。

5. for your reference：請供參考。

6. immediately：立即。亦可使用 promptly; urgently; at once; right now; without delay; as soon as possible 等。

7. upon receipt of：收到⋯時，一收⋯即⋯。

8. firm order：附有效期間的訂單。

　　文意：本公司另郵奉寄標示到舊金山運費、保險費在內貿易條件的最新價目表連同圖解目錄各乙份，謹供參考。至於裝運日期則俟收妥貴公司有效的訂單，即可着手裝運。

　　第三段提及折扣事宜。

1. little：很少的(否定用法)；a little 少量的(肯定用法)He knows little Latin, but a little French。他不懂多少拉丁文但是懂一點法文。

2. margin of profit：盈餘；賺頭(成本與售價的差額)。

3. to allow：允許，給與折扣。allowence 為津貼、折扣之意。

4. and over：以上(包括該數)與 more over 同義，以上和未滿的表達方式如下：

　　　　　100 以上 ＝above one hundred; 100 and over

　　　　　100 以上 200 未滿 ＝100 up to 200

　　　　　100 以上至 200(包括 100 和 200)

　　　　　　　＝100 through 200; 100 −200 both inclusive

　　　　　101 以上 200 未滿 ＝over 100 up to 200。

5. to secure：獲得。

6. initial order：初次訂單。

　　文意：雖然本公司開價最低因而無盈餘可圖，但是願意以百分之三的折扣優待超過一千部腳踏車的訂單，以便爭取貴公司初次訂單。

第四段樂意服務做爲結尾。

1. the opportunity of ~ing 即 the opportunity to do：有…的機會。在語意上使用動名詞比較柔和。

   出口商所謂的服務乃按訂單的指示準備貨品交貨，因此樂意提供服務具有希望獲得進口商的訂單，換言之有做生意的願望在。

文意：祈盼有機會爲貴公司服務。

四、報價(Offer)

一筆交易的成交或契約的成立必須具備兩個要件：一爲要約；二爲承諾。要約乃由當事人的一方向相對人提出一定買賣條件，表示願依此簽訂法律上有效的契約的意思表示，俗稱報價(offer)。承諾則爲當事人的另一方同意接受相對人所提出的買賣條件的意思表示，俗稱接受(acceptance)。兩個當事人中，稱前者爲報價人(offerer)而後者爲被報價人(offeree)。報價人與被報價人的意思表示必須一致，買賣契約始能成立。

由於報價可適用於買賣的任何一方，故賣方爲報價人時，此項報價爲銷貨報價(Selling offer)；而買方爲報價人，則稱購貨報價(Buying offer)，又稱出價(Bid)或訂貨(Order)。在實務上只稱 offer 時，通常係指銷貨報價而言。

·穩固報價(Firm offer)

按報價的意義，報價人須受其約束，但在實務上，有些報價預先聲明不受拘束，爲便於識別起見，前者特稱爲穩固報價(Firm offer)，即報價人在約定期間內不撤回或不變更內容的報價。穩固報價只要報價的內容並無不受拘束的聲明，firm 一字不一定須要記載，而且被報價人於約定期間(即承諾期限)內承諾，買賣契約即告成立，報價人不得悔約或要求變更契約內容。至於承諾期限是以被報價人發出承諾的時間爲準，或報價人接到該承諾的時間爲準，各國法律習慣並不一致，爲避免糾紛，報價人宜在報價中聲明承諾期限以到達報價人手中爲準。如：

We offer (firm) the following Commodities subject to your acceptance
received here by March 10, 1998,

(謹就下列貨品報價以八十七年三月十日以前接到貴公司的承諾爲有
效。)

・**未定承諾期限報價**(Free offer)

報價人未規定承諾期限，並不受任何約束的報價爲 Free offer。適用
於市場行情變動頻繁的商品的交易。報價人採用此項報價時，得任意更動
其價格或撤回報價。

・**相對報價**(Counter offer)

由於承諾必須完全符合報價人所提出的原來條件，否則不得約束報價
人。限制或變更原來條件的承諾爲有條件的承諾(Conditional acceptance)，視爲相對報價，俗稱還價(Counter offer)，不能視爲正式的承
諾。例如：

We accept your offer for 100 sets TE −5 Electronic Calculator subject
to price to be reduced to US$12. 50 C&F Los Angeles per set。

(本公司接受貴公司 100 組 TE −5 電子計算機的報價，但是以價格降
低爲每組美金十二元五角到洛杉磯運費在內價爲條件。)

相對報價在法律上屬於反要約，因此必須經對方的承諾，契約始能成
立。在貿易實務上，一筆交易往往須經過多次還價的往返，直至買賣雙方
意思合一，始能成交。

・**附條件報價**(Conditional offer)

在貿易習慣上被認爲是報價，但在法律上屬於交易的誘引者乃附條件
的要約。例如：

offer subject to our final confirmation. 須經報價人最終確認(爲條件)

的報價。

offer subject to being unsold. 未售時有效(為條件)的報價。

offer subject to prior sale. 有權先售(為條件)的報價。

offer without engagement. 不受拘束(為條件)的報價。

offer subject to market fluctuations 價格不受約束(為條件)的報價。

offer subject to approval of export license. 取得輸出許可證時有效(為條件)的報價。

offer subject to quota available. 取得配額時有效(為條件)的報價。

實踐貿易公司頃接紐約的客戶七月五日來函,欲購太陽眼鏡請求報價,主辦的林先生於是草擬了回信的要點如下:

第一段文意:謹覆七月五日大函,茲報以七月底以前接獲承諾為條件的穩固報價如下:

要點:

1.答覆來信的一般開頭用語有:

　　In reply to your letter dated……,we…

　　In response to your letter dated……,we…

　　In answer to your letter dated……,we…

　　若欲表達遵照來函已……,則可用:

　　As requested in your letter dated……,we…

　　As instructed in your letter dated……,we…

　　或 In accordance with the request in your letter dated……,we…

　　In compliance with the instructions in your letter dated……,we…

2.穩固報價:to offer you firm 此際 offer 為動詞,firm 為副詞用法。

3.法律上表示條件的慣用語為 subject to(以 ~為條件的)。

4.……之前;……以前,用 by(=not later than)表達,而報價使用的 by 通常指營業時間(business hours)。

　　故七月底以前為:by the end of July, 1998 或 by July 31, 1998。

5.七月底以前接獲承諾為條件:subject to your acceptance reached

here by July 31, 1998.

6.如下：as follows, follow 為非人稱動詞，不論其主詞的數，均使用第三人稱單數形。

His opinions were as follows 他的見解如下。

文稿：In response to your letter dated July 5, 1998, we offer you firm subject to your acceptance received here by July 31, 1998 as follows:

第二段：本案的交易內容為爽克牌 A 型太陽眼鏡 150 打，以紐約港起岸價格計價每打美金 70 元，九月份交貨，憑不可撤銷的即期信用狀付款，林先生將報價的內容表列如下：

1.品名：太陽眼鏡　　　　　Article: Sunglasses

2.品質：爽克牌 A 型　　　　Quality: " Suncool " Brand Model A

3.數量：150 打　　　　　　Quantity: 150 dozen

4.價格：紐約起岸價格每打美金 70 元。

　Price: US$70. 00 per doz. CIF New York

5.交貨：九月份　　　　　　Shipment: September

對外貿易的交貨以裝運日為準，而其期間採月份者較多，故 September Shipment 係指九月一日起至九月三十日為止的期間內裝運即可。而 September / October 則指九月一日起至十月三十一日為止的裝運期間。

6.付款：憑不可撤銷的即期信用狀付款。在貿易的各項交易條件 (Terms & Conditions) 中 terms 一詞專指付款條件或費用、手續費、佣金等有關金錢的條款，故常被用來代 Terms of Payment。憑信用狀付款，就出口商的立場而言，係出票行為，故寫作 Draft at sight 或 sight draft；不可撤銷信用狀為 Irrevocable L / C；而匯票是憑信用狀開立者，換言之在信用狀項下開立的，故用 under，於是付款條件為：

Terms: Sight Draft under Irrevocable L / C.

文稿：Article: Sunglasses.

Quality: "Suncool" Brand Model A.

Quantity: 150 dozen.

Price: US$70.00 per doz. CIF New York.

Shipment: September.

Terms: Sight Draft under Irrevocable Letter of Credit.

第三段：為了表示上述是往後不再有的最好的報價，林先生略提業務及市場狀況，以便爭取對方迅速承諾報價。

文意：此乃本公司所能給予而且將不再重報的最好報價，目前接獲來自貴國的大量訂單，而且行情堅挺，在這種情況下有各種跡象顯示不久價格將上漲。

1.本公司所能給予的最佳報價：

This is the best offer we can make.

2.將不再重報的報價＝本公司將來不再報同樣的價：

We cannot repeat this kind of offer in the future.

惟配合前半句以報價為主詞，故宜改用被動語態：

This kind of offer cannot be repeated in the future.

把 1.2.連起來 This is the best offer we can make and this kind of offer cannot be repeated in the future.

3.接獲大量訂單的基本語態為 We obtain large orders，表達目前的情況則用進行式 We are obtaining large orders now 本敘述宜改為被動語態以便減少「你、我」的字眼在信中出現的機會，同時用動詞 rush in (湧到、衝進) 更能形容大量訂單，故貴國的訂單大量湧到可寫作：Now large orders are rushing in from your country.

4.行市堅挺：the market here is very strong

5.在此種情況下：Under the circumstances

6.各種跡象：every indication；樣樣都在暗示，暗示甚麼呢？

7.價格即將上漲就是暗示的內容，故以 that 連接而成。

that the prices will rise soon.

把 5. 6. 7.連起來：under the circumstances, there is every indication that the prices will rise soon.

第四段：以邀請儘速接受報價作為結尾。

文意：為此，敬請貴公司迅速接受報價是幸。

1.敬請 to advise，由於價格不久就要上漲，事不可猶豫，因而用奉勸的字樣提醒對方。

2.奉勸貴公司接受報價：We advise you to accept this offer。

此際以 therefore 插入更能圓腔。

We advise you, therefore, to accept this offer.

3.儘速：immediately, soon, at once 這些字都含有迅速之意，但太過於直接，不如 without loss of time（不浪費時間＝不猶豫）火急中不失文雅。

林先生繕妥的全文如下〔例 6 －3 〕：

〔例 6 – 3 〕

Gentlemen:

In response to your letter dated July 5, 1998, we offer you firm subject to your acceptance received here by July 31, 1998 as follows:

        Article: Sunglasses
        Quality: " Suncool " Brand Model A
        Quantity: 150 dozen
        Price: US$70. 00 per doz. CIF New York
        Shipment: September
        Terms: Sight Draft under Irrevocable L / C

This is the best offer we can make and this kind of offer cannot be repeated in the future. Now large orders are rushing in from your country and the market here is very strong. Under the circumstances, there is every indication that the prices will rise soon.

We advise you, therefore, to accept this offer without loss of time.

                            Yours truly,

五、接受報價（Accepting an Offer）

　　實踐貿易公司七月初寄發的外銷太陽眼鏡的報價函於下旬收到了對方的回信〔例6-4〕接受了報價。

　　第一段：

　　1. to accept：常用於對外貿易的各種場合，其意義及用例如下：

　　　　to accept your offer：承諾報價。

　　　　to accept your payment：領受給付。

　　　　to accept your draft：承兌匯票。

　　　　回信時不必一一複述案由，只要提及對方來信日期即可，但是本函卻複述報價的主要內容，此乃報價須受其約束，複述可以確認報價內容。

　　文意：本公司茲承諾貴公司七月十五日函，有關爽克牌 A 型太陽眼鏡 150 打，紐約港起岸價格每打美金 70 元，九月份交貨的報價。

　　第二段：

　　1. herewith：附此；同此。We are sending you herewith … 隨函檢送。

　　2. Purchase Note：購貨單；購貨合約。交易成交後需要簽訂買賣契約，規定有關交易內容及條件，此項契約由買方繕製者即為 Purchase Note 或 Purchase Contract。通常製一式兩份經簽署後寄給賣方，請求賣方簽署後抽存一份，另一份寄回買方收執。

　　文意：隨函檢送第七〇五號購貨單，以確認本次交易。

　　第三段：交易成交後，賣方有交貨的義務，買方則負付款的義務。憑信用狀為付款條件的交易，應由買方先履行義務，即依合約的規定開發信用狀，而後由賣方履行交貨的義務。在這一段買方表示已申請開發信用狀，好讓賣方開始準備交貨的工作。

　　1. to cover：補償；抵付；擔保；投保。

第六章　交易的進行

〔例 6 − 4 〕

Gentlemen:

We are glad to accept your offer dated July 15, 1998 for 150 dozen Sunglasses " Suncool " Model A at US$70.00 per dozen CIF New York for September shipment.

To confirm this transaction, we are sending you here with our Purchase Note No. 705.

In order to cover the amount of this purchase, we have arranged with our bankers for an Irrevocable Letter of Credit to be opened in your favor.

Since this transaction is very important to us mutually, we would like you to give it your best attention to satisfy us in every respect.

Yours sincerely,

Encl.

2. amount：金額；合計；總數；與 sum total, aggregate 同義。

3. to arrange with…for…：爲…(事宜)與…(人)洽商；向…申請…。

4. in your favor：以貴公司爲受益人；貴公司抬頭的。

文意：茲爲償付本批貨款，本公司業已向往來銀行申請開發貴公司爲受益人之不可撤銷信用狀。

第四段：叮嚀對方妥善交貨。

1. transaction：交易；買賣；與 business, dealings, sales, trade 同義。

2. mutually：互相地；彼此共同。

3. would like：願意；但願。

4. best attention：專心；特別留意。

5. in every respect：各方面；respect 爲細節，有關方面；in respect of

則爲關於……。

　　文意：鑒於本案極受貴我雙方的重視，祈盼貴公司在各方面特予費神，務請盡善盡美。

六、還價（Counter Offer）

　　實踐貿易公司五月間外銷美國女用套頭上衣的報價遭到美商的還價〔例6－5〕。還價是有條件的承諾，因此信中必須具體地說明不接受的原因，並提出建議或辦法等徵求對方的同意。

　　第一段：由於還價是針對不接受的條件提出相對的報價，故對原報價仍宜先致謝。

1. offer dated May 6, 1998 on…：八十七年五月六日有關…的報價。
2. V－neck Pullover：V領套頭上衣。

文意：五月六日大函有關女用V領套頭上衣報價案敬悉。

第二段：解釋原報價無法接受的具體理由。

1. however：然而，頓一下或轉換語氣常用的連接詞。
2. We are not in a position to accept…本公司歉難接受…，貿易書信中既具體又客氣的婉拒語句。報價非能不能接受的問題，而是條件合不合意的問題，若用 We can not accept 則語氣硬得似乎無法妥協。
3. since：因爲。because 是對方不知而想知的理由和原因；
   since 則與 as 一樣，所敘述的理由是雙方所明瞭者。
4. prospective buyers：未來的買方，欲訂貨者。
5. another source＝another channel：另一途徑，指本案報價人以外的其他業主。
6. attractive offer：吸引人的報價；低廉的報價（offer at attractive prices）。

第六章　交易的進行

〔例 6－5〕

Gentlemen:

Thank you for your offer dated May 6, 1998 on Ladies V－neck Pullover.

However, we are not in a position to accept your offer since our prospective buyers state that they have received from another source in Taiwan a much attractive offer which is around 8% below your price.

In case your price will be reduced to US$75. 48 per doz. FOB Taiwan port, we may be able to obtain a substantial orders from them. Please bear in mind, in this connection, that pullover is one of the items that our buyer regularly needs, and that we could expect to secure a foothold in this line.

Your favorable reply will be appreciated.

Very truly yours,

---

7. around：大約，……左右(＝about)

文意：鑒於本公司的訂戶，自台灣其他業主取得便宜百分之八左右的報價，貴公司的報價歉難接受。

第三段：根據上述不接受的具體理由，計算出可行的價格作為反要約。由於反要約的內容多屬價格問題，此乃反要約之所以被稱為還價的原由。

1. in case：若，如果。

2. to be reduced to…：減為……，US$50. 00 改為 US$45. 00 的表達方式有二：

(1) to be reduced to US$45. 00 減為美金四十五元(以總價表達)。

(2) to be reduced by US$5. 00 減少美金五元(以差價表達)。

3. substantial orders：大量的訂單；數量可觀的訂單。

4. in this connection：指這次的交易。connection 是業務上的關係。

5. to secure a foothold：鞏固地盤。foothold 立足點。

文意：若貴公司的價格能減為每打離岸價格美金七十五元四角八分，則可望取得可觀數量的訂單。套頭衣是本公司客戶經常採購的項目之一；同時本公司藉此得以鞏固業務地盤，這兩點尚請貴公司明察。

第四段：希望對方答應減價。

文意：敬候佳音。

練習問題

一、試述如何爭取國外進口商的詢價函？

二、試述報價時所提的各項交易條件。

三、我國進出口貨品分為那幾類？進出口廠商為何須認識其經手貨品之類別？

四、貨品的品質如何表示？最常見者為何？

五、完整的價格表達應包括那幾項內容？

六、外匯的匯率分幾種？

七、全程均以海運交貨且以輸出地交貨條件下所報的貿易條件有幾種？若改用航空或複合運輸，則適用何種貿易條件，試列舉之。

八、試述對外貿易的付款方式。

九、試述海上保險的基本險和常見的附加險。

十、〔例 6－2〕的開頭部分過於冗長，不符簡潔之原則，試更改之。

十一、試根據下列資料，填製報價單：

　　(1)被報價人：New York Trading Co., Ltd., World Trade Center, New York, N. Y.,U. S. A.

　　(2)報價地點、日期：台北，民國 87 年 6 月 18 日

　　(3)報價單編號：SCTD－12345

　　(4)商品名稱：白雪牌滑石粉 (Talc Powder, Snow White Brand)

　　(5)品質規格：白度 90% 以上，作為化粧 (Cosmetic) 之用。篩孔

(mesh)：325

(6)數量：500 公噸，得有 10% 的寬限。

(7)價格：基隆船上交貨每公噸 83 美元。

(8)包裝：以三層牛皮紙袋包裝，每袋淨重 50 公斤，中層裏 (Middle Ply Coated with) 防水布 (Tarpaulin) 以爲防水之用。

(9)付款：憑第一流銀行開發以賣方爲受益人之不可撤銷、可轉讓、即期信用狀且須在接受後卅天內開達賣方。

(10)交貨：收到正本信用狀後 60 天內裝運，允許分批裝運，允許轉運。

(11)保險：由買方自理。

(12)檢驗：以製造廠之檢驗證明爲準，買方如有要求其他任何檢驗，由買方付費。

(13)匯率風險：如有任何匯率變動，其風險由買方頁擔。

(14)有效期限：本報價有效至台北時間，民國 87 年 7 月 18 日。

十二、對本公司二月十五日的報價，國外的進口商於三月二日來信，認爲較其他公司所報價格高約 3%。根據調查目前的運費、保險費並無差額費率。則台灣銷往該國的同類貨色均高 3%，此乃不可思議。莫非此乃對方殺價的藉口，爲爭取這筆訂單，決定降價 1% 賣出。請你草擬回信。

Trade 第七章 English

交易的成立

在實務上，出口商作報價之前，不但考慮其出口成本、費用及利潤，而且尚需顧及到進口地的市場行情，因此除非品質不符合價格或價格太離譜，通常經過幾次彼此討價還價的結果，雙方的意思漸趨一致，最後達成交易。

一、訂貨信（Order Letter）

交易一旦成立，買賣雙方必須以書面確認有關的交易內容，作為日後履約的依據。訂貨單又稱訂單，乃確認交易最簡單的方式，實務上，採互相確認的方式，即由買方寄送訂單並由賣方覆以確認訂單（Acknowledgement of Order）。訂單述及訂貨的意思表示、訂貨的依據（例如目錄或價目表等），並詳載訂貨有關的各項內容，此項內容類似報價的內容，但是在報價時常被省略的包裝和保險條款，在訂單裏因與交貨的安全與風險的保障有密切關係，必須叮嚀清楚。

若訂單的內容繁多複雜，則可另備專用的訂單（Order Sheet）並附上信函郵寄賣方。一般的進口廠商均備有特定項目的訂單格式（Order Form）或契約書（Contract Form），經辦人只需繕打細節即可。購貨單（Purchase Note）亦可代替訂單。訂單除了於接受報價時由買方寄發，如〔例 6 −4〕之外，亦有買方憑賣方所送貨樣、價目表等函請訂購者。

實踐貿易公司外銷美國套頭上衣案以美金 75.48 成交，美商來函訂貨〔例 7 −1〕。

第一段：美商使用格式的訂單，開端是標準的套句。

1. to place the following order with you：向貴公司訂購下列的貨品。

2. to set forth：宣佈，記載。

文意：敬覆五月廿三日大函，茲依照下述各項條件向貴公司訂貨。

第二段：訂單的內容類似於報價的內容，而且所列項目較之報價時為多，例如包裝和刷嘜等條件甚少列在報價單上，但是訂單上則少不了這兩項。本案除了這兩項條件之外，買方特別要求貨品的配色。

〔例 7 – 1〕

Gentlemen:

Re Order No. 817583

In reply to your letter dated May 23, 1998, we are glad to place the following order with you on the terms and conditions as set forth below.

Commodities: Ladies′ V – neck pullover
Quality: same as sample No. ST – 05
Quantity: 80 dozen
Price: US$75. 48 per doz. FOB Taiwan port
Amount: US$6, 038. 40
Shipment: July, 1998
Packing: One dozen in a carton, 10 carton in a wooden case
Marks:

New York
C / 1 – 8

Payment: by Irrevocable Sight L / C
Insurance: covered by ourselves
Remarks: the assortment should be as follows:

	S	M	L
White	4	8	8
Red	5	10	15
Blue	6	12	12

We have just arranged with J. Henry Schroder Bank and Trust Co., New York for an Irrevocable Letter of Credit to be opened in your favor.

Your usual best attention to the execution of the present order so as to satisfy us in every respect will be much appreciated.

Yours faithfully,

1. Sample No. ST –05：第 ST –05 號樣品。憑樣品買賣的交易以樣品代號或編號作為品質的標準。

2. Carton：紙板盒。

3. Wooden Case：木箱。

4. to cover：投保。

5. remark：備註；注意事項。

6. assortment：貨品的配色。本案的套頭上衣分白、紅、藍三色，各種顏色所需件數，按上衣的大(L)、中(M)、小(S)尺寸配件，亦稱爲 color and size assortment。

文意：貨品：女用 V 領套頭上衣。

　　　品質：以 ST−05 號樣品爲準。

　　　數量：八十打。

　　　價格：台灣港口離岸價格每打美金 75. 48 元。

　　　交貨：七月內。

　　　包裝：紙板盒一打裝，十盒裝一木箱。

　　　刷嘜：　　　◇AMC◇

　　　　　　　　　New York

　　　　　　　　　C / 1 −8

　　　付款：憑不可撤銷的即期信用狀。

　　　保險：本公司自行投保。

　　　備註：務請照下列的配色數量交貨。

	小號	中號	大號
白色	4	8	8
紅色	5	10	15
藍色	6	12	12

第三段：進口商希望出口商早日交貨，則自己亦必須儘早履行義務，即開發信用狀。開發信用狀的目的除了履行進口商的義務之外，亦得因此要求出口商照信用狀條款交貨。

1. J. Henry Schroder Bank and Trust Co., New York：紐約的傑亨修銀行，係本案美國廠商的往來銀行，也是信用狀的開狀銀行。

2. in your favor：以貴公司爲受益人。

文意：本公司已洽請紐約的傑亨修銀行，開發以貴公司爲受益人的不可撤銷即期信用狀。

第四段：在訂單的結尾，進口商不外多叮嚀出口商費神交貨。

1. best attention：留神，最大的注意力。

2. execution：執行、完成。execution of the order 處理訂單，指交貨。

文意：敬請貴公司照往例費神交貨，並盼各方面均能令人滿意是幸。

二、銷貨確認（Sales Confirmation）

對於進口商的訂單，爲確認交易成立，出口商往往覆信確認訂單。實踐貿易公司接獲美國訂購女用 V 領套頭上衣的訂單〔例 7－1〕之後，草擬回信如下：

第一段重點：確認六月二日 AI－07 號訂單。

1. 「確認」用 to acknowledge，並套用圓腔的詞句而成 We have the pleasure of acknowledging…。

2. 訂單的內容由於另附銷貨單（Sales Note）詳述，此處僅列訂貨數量（80 dozen）、品名（Ladies' V－neck Pullover）、價格（at US$75. 48 FOB Taiwan Port）及交貨期（（July shipment）即可。

文稿：We have the pleasure of acknowledging your order No. AI－07 dated June 2, 1998 for 80 dozen Ladies, V－neck Pullover @ US $75. 48 FOB Taiwan Port for July Shipment.

文意：本公司茲確認貴公司六月二日 AI－07 號訂購 80 打之女用 V 領套頭上衣七月份交貨台灣港口離岸價格美金七十五元四角八分的訂單。

第二段重點：檢送銷貨單，請查收。

1. 隨函檢送銷貨單（SE－115），爲表達 you 口氣，可利用 you will find enclosed our Sales Note No. SE－115.

2. 其內容悉照訂單。「悉照訂單內容」其意義乃各方面都無錯誤；換言之，銷貨單在各方面都是符合的，Our sales Note No. SE −115 which is correct in all respects.，而本公司有把握一切內容都符合，即本公司相信經貴公司查對不致有不符的內容，We trust you will find correct in all respects. 。

文稿：You will find enclosed our Sales Note No. SE −115 which we trust you will find correct in all respects.

文意：隨函檢送 SE −115 號銷貨單，諒一切合符尊意。

第三段重點：遵照指示早日交貨。

1.遵照貴公司的要求 as you requested, we……

2.本公司在盡力 We are now doing our best.

3.交貨 to ship your order

4.早日，無法確定日期，故用儘早的字眼 at the earliest possible date.

文稿：As you requested, we are now doing our best to ship your order at the earliest possible date.

文意：遵照貴公司的要求，本公司當儘力早日交貨。

第四段重點：謝謝訂購，並盼繼續惠顧。

1.謝謝本次訂單：Many thanks for this order.

2.請繼續惠顧：繼續惠顧＝繼續訂貨＝續訂 repeat orders 諒可獲取續訂：We may have repeat orders from you.

3. repeat orders 是續訂，有關訂單的名稱尚有 initial order 首次訂單；sample order 樣品訂單；telephone order 電話訂貨；cable order 電報訂貨。有關訂單的動詞則有：

　　to place an order with：向…訂貨，向…發訂單

　　to carry out (fulfill, execute) order：處理訂單；進行交貨

　　to decline order： 婉謝訂貨。

4.為了爭取續訂應提出充分的理由，本公司的理由乃深具信心妥善處理訂單。

　　深具信心⇒有把握⇒向貴公司擔保 We assure you；擔保的事項

「妥善處理訂單」為 our best attention to its execution；於是整句的意思為 We assure you of our best attention to its execution.

5.根據上述理由盼能再爭取訂單，故寫作：so that we may have repeat orders from you.

文稿：Many thanks for this order and we assure you of our best attention to its execution so that we may have repeat orders from you.

文意：此次承蒙訂貨不勝感激，當盡力辦妥交貨，企盼惠予續訂是幸。

以上的文稿繕妥後，即為對進口商訂單的確認〔例7-2〕，至於附件的銷貨確認函的格式，請參考第八章練習問題八。

〔例7-2〕

Gentlemen:

We have the pleasure of acknowledging your order No. AI-07 dated June 2, 1998 for 80 dozen Ladies' V-neck Pullover @ US$75.48 FOB Taiwan port for July shipment.

You will find enclosed our Sales Note No. SE-115 which we trust you will find correct in all respects.

As you requested, we are now doing our best to ship your order at the earliest possible date.

Many thanks for this order and we assure you of our best attention to its execution so that we may have repeat orders from you.

                                        Yours very truly,

Encl.

三、訂約(Concluding a Contract)

訂單一經賣方確認之後，交易即告正式成交，自此買賣雙方均負有履約的義務。為恐口說無憑，將雙方的權利義務訂明於文書，經雙方簽署以為履約的主要依據者乃買賣契約。簽訂契約(Concluding a Contract 或 Entering into an agreement)有政府或貿易團體之間經過協議而簽訂的貿易協定(Trade Agreement)及進出口廠商之間為大筆正式簽訂的契約；由出口商主動制定者為銷售契約(Sales Contract)，進口商主動制定者為購買契約(Purchase Contract)。實務上，由於一般進出口廠商之間的交易金額不大或交貨的期限不甚長，更由於不便為每筆交易雙方聚首簽訂契約，故正式簽訂契約者少而採換文確認者多。由賣方出具的確認函有銷貨單(Sales Note)、銷貨確認函(Sales Confirmation or Confirmation of Sales)、訂貨確認書(Confirmation of Order 或 Acknowledgement of Order)等。由買方出具者有購貨單(Purchase Note)和購貨確認函(Purchase Confirmation 或 Confirmation of Purchase)。〔例 7−2〕即為一般常見的銷貨確認函。

正式簽訂的買賣契約，其內容由一方製作一式兩份(in duplicate)，經雙方當面認可後簽署分別留存一份，或由製作的一方簽署兩份逐寄對方，由對方簽署後保留正本，副本則寄還製作的一方。常見的簽訂條款如下：

1. Commodity　　　　　　　　（商品條款）
2. Specifications　　　　　　　（規格條款）
3. Quality　　　　　　　　　　（品質條款）
4. Quantity　　　　　　　　　（數量條款）
5. Price　　　　　　　　　　　（價格條款）
6. Payment　　　　　　　　　（付款條款）
7. Packing　　　　　　　　　（包裝條款）
8. Shipping marks　　　　　　（刷嘜條款）
9. Shipment　　　　　　　　　（裝運條款）

10. Loading and Discharging	(裝卸貨條款)
11. Insurance	(保險條款)
12. Inspection	(檢驗或公證條款)
13. Export or Import Licence	(輸出入許可證條款)
14. Taxes and duties	(稅捐條款)
15. Exchange risk, Insurance risk, Freight risk	(匯率、保險、運費等風險條款)
16. Claim	(索賠條款)
17. Force Majeure	(不可抗力條款)
18. Arbitration	(仲裁條款)
19. Governing Laws (or regulation)	(準據法條款)
20. Validity	(有效期限條款)

　　實踐貿易公司外銷女用 V 領套頭上衣乙案採用訂貨確認書,係屬於換文確認的訂約方式,若欲作成正式的買賣契約,可參照下述步驟製作:

第一步驟:標明合約名稱及當事人

Contract for purchase and sale of Ladies, V—neck Pullover American Market Corporation, New York (hereinafter called the Buyer) has agreed to buy, and Shih Chien Trading Co., Ltd., Taipei (hereinafter called the Seller) has agreed to sell Ladies, V—neck Pullover on the following Terms and Conditions:

1. hereinafter called:…以下簡稱為…

文意:女用套頭上衣買賣契約。

　　紐約的美國市場公司(以下簡稱為買方)和台北的實踐貿易公司(以下簡稱為賣方)雙方茲同意照以下各項條件買賣女用套頭上衣。

第二步驟:將訂單內容依序陳述,參閱〔例 7−1 〕。

　　I. Commodities: Ladies, V—neck Pullover

　　II. Quality: The seller should supply the commodities exactly the same as sample No. ST −05 on which Buyer approved this order.

Ⅲ. Quantity: 80 dozen (Eighty Dozen)

Ⅳ. Price: US$75. 48 per dozen FOB Taiwan port.

Ⅴ. Total Amount: US$6, 038. 40 (US DOLLARS SIX THOUSAND THIRTY EIGHT AND CENTS FORTY ONLY)

Ⅵ. Shipment: The shipment shall be effected within July 1, 1998 to July 31, 1998 in one shipment.

Ⅶ. Packing: Each dozen of the contracted pullover shall be packed in a carton and 10 cartons in a wooden case.

Ⅷ. Marking: The following marks is to be marked on each wooden case.

<AMC>

New York

C / 1 −8

Ⅸ. Payment: An Irrevocable sight Letter of Credit for full contract amount in favor of Seller is to be opend by the Buyer not later than June 30, 1998. Seller may draw a draft under the credit to cover full invoice value.

Ⅹ. Insurance: Insurance shall be covered by the Buyer.

文意： I. 貨品：女用 V 領套頭上衣。

　　 Ⅱ. 品質：賣方所交貨品其品質務必與買方憑以訂貨之 ST −05 號樣品完全相同。

　　 Ⅲ. 數量：八十打。

　　 Ⅳ. 價格：台灣港口離岸價格每打美金七十五元四角八分。

　　 Ⅴ. 總金額：美金陸仟零叄拾捌元肆角正。

　　 Ⅵ. 裝運：八十七年七月一日起至八十七年七月卅一日之間一次裝運。

Ⅶ. 包裝：每打裝一紙板盒，每十個紙板盒裝成一木箱。

Ⅷ. 刷嘜：每一木箱的刷嘜如下：

New York

C／1 −8

Ⅸ. 付款：買方須於八十七年六月三十日之前，開發契約金額
全額，以賣方爲受益人的不可撤銷即期信用狀，賣
方得在信用狀項下出票抵償發票金額。

Ⅹ. 保險：保險由買方自行投保。

第三步驟：加列特別條款。本案有關貨品的配色、品質的檢驗、輸出許可
證及不可抗力等乃值得列入本契約條款。

Ⅺ. 　　Assortment: The Seller shall make the shipment according to
the following assortment.

	S	M	L
White	4	8	8
Red	5	10	15
Blue	6	12	12

Ⅻ. Inspection: Specification, Quality, Quantity, Packing and Weight of
the Commodity shall be inspected prior to shipment by
an independent public surveyor. The inspector's certificate to this effect shall be final. Inspection fee shall be
for Seller's account.

ⅩⅢ. Export License: The fulfilment of this contract is subject to export
license, if necessary, granted by BOFT of the
Seller.

ⅩⅣ. Force Majeure: The Seller is not responsible for delay or non −
performance of his contractual obligation to sell,
and the Buyer, is not responsible for delay or

non‑performance of its contractual obligation to purchase all or any part of the supplies caused by war, blockade, revolution, insurrection, civil commotions, riots, mobilizations, strikes, lock‑outs, acts of God, plague or other epidemic, fire, flood, obstruction of navigation by ice at port of delivery and destination, acts of government or public enemy.

文意：XI. 貨品的配色：賣方須按下列配色交貨。

	小	中	大
白色	4	8	8
紅色	5	10	15
藍色	6	12	12

XII. 檢驗：貨品的規格、品質、數量、包裝及重量須於裝運之前由獨立公證行檢驗並憑該公證行的檢驗證明為準。檢驗費用由賣方負擔。

XIII. 輸出許可證：本契約的履行若有必要則以賣方取得國際貿易局的輸出許可證為條件。

XIV. 不可抗力條款：若因戰爭、封鎖、革命、叛亂、民變、暴動、動員、罷工、停工、天災、疫病或其他傳染病、火災、洪水、裝貨及目的地之港口冰凍阻塞，政令或敵僑之行為等原因導致賣方遲延或未能履行契約上出售的義務，而買方遲延或未能履行契約上購買的義務時，賣方或買方對此並不必負責。

第四步驟：可能的話，加列仲裁條款並打出買賣雙方名稱以便簽署，合約即訂妥。

XV. Arbitration: All dispute arising in connection with the present contract shall be finally settled under the

Rules of Commercial Arbitration Association of Republic of China, and the award to be final and binding on both parties.

文意：ⅩⅤ. 仲裁條款：所有與本合約有關之爭執，應依中華民國仲裁協會之規則謀求最後之解決並以仲裁人之判決爲準，買賣雙方均不得異議。

以上的合約內容繕妥後即爲〔例 7－3〕。

〔例 7－3〕

Contract No. EI－032   Date: June 10, 1998

Contract for Purchase and
Sale of Ladies' V－neck Pullover

American Market Corporation, New York (Hereinafter called the Buyer) has agreed to buy, and Shih Chien Trading Co., Ltd., Taipei (Hereinafter called the Seller) has agreed to sell Ladies' V－neck Pullover on the following terms and conditions:

Ⅰ. Commodities: Ladies' V－neck Pullover.
Ⅱ. Quality: The Seller should supply the commodities exactly the same as sample No. ST－05 on which Buyer approved this order.
Ⅲ. Quantity: 80 dozen (eighty dozen).
Ⅳ. Price: US$75.48 per dozen FOB Taiwan Port.
Ⅴ. Total Amount: US$6,038.40 (US Dollars Six Thousand Thirty Eight and Cents Forty Only)
Ⅵ. Shipment: The shipment shall be effected within July 1, 1998 to July 31, 1998 in one shipment.
Ⅶ. Packing: Each dozen of the contracted pullover shall be packed in a carton and 10 cartons in a wooden case.
Ⅷ. Marking: The following marks is to be marked on each wooden case.

New York
C／1－8

Ⅸ. Payment: An Irrevocable sight Letter of Credit for full contract amount in favor of Seller is to be opened by the Buyer not later than June 30, 1998. Seller may draw a draft under the credit to cover full invoice value.

Ⅹ. Insurance: Insurance shall be covered by the Buyer.
Ⅺ. Assortment: The Seller shall make the shipment according to the following assortment.

	S	M	L
white	4	8	8
red	5	10	15
blue	6	12	12

Ⅻ. Inspection: Specification, quality, quantity, packing and weight of the commodity shall be inspected prior to shipment by an independent public surveyor. The inspector's certificate to this effect shall be final. Inspection fee shall be for Seller's account.

XIII. Export License: The fulfilment of this contract is subject to export license, if necessary, granted by BOFT of the Seller.

XIV. Force Majeure: The Seller is not responsible for delay or non-performance of his contractual obligation to sell, and the Buyer, is not responsible for delay or non-performance of its contractual obligation to purchase all or any part of the supplies caused by war, blockade, revolution, insurrection, civil commotions, riots, mobilizations, strikes, lockouts, acts of God, plague or other epidemic, fire, flood, obstruction of navigation by ice at port of delivery and destination, acts of government or public enemy.

XV. Arbitration: All disputes arising in connection with the present contract shall be finally settled under the Rules of Commercial Arbitration Association of Republic of China, and the award to be final and binding on both parties.

Buyer: American Market Corp., New York
(signature)
Seller: Shih Chien Trading Co., Ltd., Taipei
(Signature)

練習問題

一、試述接受報價的意義及其要件和方式。
二、試述訂單的內容和報價單有何不同。
三、試就下列有關訂貨的詞語，簡述其意義並說明其用法。

　　　　　initial order

limited order

market order

open order

original order

repeat order

sample order

sizeable order

split order

substantial order

telephone order

trial order

verbal order

order blank

order book

order form

order memo

order number

order sheet

to accept an order

to book an order

to cancel an order

to confirm an order

to countermand an order

to execute an order

to get [or obtain] an order

to mail an order

to place an order

to repeat an order

to secure an order

四、試根據下列資料填製一份購貨確認書或銷貨確認書，格式參照第八章
練習問題八。

(1)賣方 IDA Trading Company；買方 ARAB Trading Company

(2)確認日期：民國 87 年 9 月 15 日。

(3)商品：桌上電扇。

(4)規格：16 吋綠色，其餘如賣方第 100 號型錄說明。

(5)價格：至吉達港運費、保險費及佣金百分之五在內價，每台美金十
五元正。

(6)數量：五百台，但賣方得多裝或少裝百分之十。

(7)包裝：出口標準木箱，箍以鐵皮條 (Iron strap)。

(8)保險：按發票金額加百分之十投保水漬險及竊盜及短卸險、兵險。

(9)交貨：收到正本信用狀後六十天內交貨，准許分批交貨，不可轉
運。

(10)付款方式：憑賣方為受益人的一流銀行所開發不可撤銷、可轉讓信
用狀，以見票即付匯票領款。

(11)匯率變動風險由買賣雙方平均負擔。

(12)嘜頭及件號：ARAB 吉達，件號自第 1 號起。

五、詳讀〔例 7—4〕的合約並回答各項問題：

1.本合約的當事人是誰？

2.為何簽訂本合約？

3.合約由誰主稿，換言之，係何種合約？

4.本合約的主要條款有幾項？試指出條款名稱。

5.合約與銷貨確認函有何不同？與訂單又有何區別？

〔例 7 －4 〕

EXPORT CONTRACT   CONTRACT No. 12345
                               Date: Jan. 10, 1998

U. S. Soybean Corp., New York. N. Y., U. S. A. (Hereinafter referred to as
SELLER) agree to sell and Taiwan Trading Co., Ltd., Taipei, Taiwan (Hereinafter
referred to as BUYER) agree to buy the following Commodity according to the
terms and conditions set forth below:

Commodity   : U. S. Yellow Soybean, Grade No. 2.

Specifications   : 1. Test weight per bushel: 54 1bs, min.
                    2. Moistrue: 14% max.
                    3. Splits: 20% max.
                    4. Total demaged kernels: 3% max. (including heat demaged
                       kernels: 0. 5% max.)
                    5. Foreign material: 2% max.
                    6. Brown, Black, and / or Bicolored soybeans: 2% max.

Quantity   : 3,000 metric tons.

Price   : U. S. $ 215 per metric ton net C & F Keelung, berth terms.

Shipment   : To be effected not later than Jan. 21, 1998 by direct liner.

Payment   : By Irrevocable letter of credit available by sight drafts accom-
            panied by the following documents:
            1. Clean on Board Ocean Bill of Lading.
            2. Commercial Invoice.
            3. Inspection Certificate issued by the U. S. Government.
            4. Weight Certificate issued by an independent licensed weigher.
            5. Official Phytosanitary Certificate issued by the U. S. Gov-
              ernment.

Insurance   : Tobe covered by the Buyer.

Inspection&   : The following certificates covering quality and quantity at Seller's
Weighing   cost are required and shall be considered as final.
            1) Quality inspection at the loading port shall be made by the
              U. S. Dept. of Agriculture under the U. S. Grain Standards
              Act, and a Federal Appeal Grade Certificate certifying that the
              quality of the soybeans strictly to the specification of U. S.
              Yellow Soybean Grade No. 2 or better shall be produced.
            2) Inspection on weight at the loading port shall be made by
              an independent licensed official weigher and an official
              weight certificate shall be produced.
            3) An official phytosnaitary certificate shall be issued by the
              U. S. Government.

Arbitration   : Any dispute be referred immediately to the Arbitration Asso-
            ciation of the New York, N. Y.

          Seller: Buyer:

       _____   _____

第八章 Trade English

交易的履行

一、信用狀（Letter of Credit）

買賣契約以信用狀為付款條件者，站在買方立場的進口商應履行開發信用狀的義務，而賣方的出口商憑此履行交貨的義務並取得貨款。信用狀乃目前對外貿易最常見的支付貨款的工具（Payment Instrument），係屬於出票的方式（參閱第六章）。根據國際商會一九九三年修訂的信用狀統一慣例（Uniform Customs and Practice for Documentary Credits 1993 Revision, Publication No. 500）之第二條：「跟單信用狀及擔保信用狀係指銀行為本身或循客戶（即申請人）之請求，照其指示所做之安排（Arrangement）」。憑規定且符合信用狀條款之單據，(1)對受益人或其指定人辦理付款，或對受益人所開匯票辦理承兌並予付款，或(2)授權另一銀行辦理上項付款，或對上項匯票辦理承兌並予付款，或(3)授權另一銀行辦理讓購。簡言之，信用狀為開狀銀行對於信用狀之受益人所提示，符合信用狀條款之單據或跟單匯票履行承兌並付款或讓購的承諾書信。

假設對外貿易不以信用狀為償付貨款的工具，則出口商有兩個問題待解：

第一：由於出口商無法隨時知悉進口商目前的營運情況，縱然憑過去往來的紀錄或從同業間的傳聞，甚至透過徵信機構的資料可佐推定其信用狀況，但是每當交貨完畢都得逐筆擔心進口商是否會如約償付貨款。因為進口商可能藉故，如物價下跌，另訂條件優厚的買賣，或資金不足為由片面毀約，甚至到貨後拒絕提貨並要求折讓等。

第二：在沒有信用狀的情況下，出口商收回貨款的方法不外(1)要求進口商匯付，(2)要求銀行貼現，(3)委由銀行託收。無論採取其中任一方法，均有不利於出口商之處。蓋(1)要求進口商匯付貨款，等於將信用風險及融資上的負擔轉嫁，往往不輕易為進口商所接受。(2)出口商出具以進口商為被出票人（Drawee）的匯票連同跟單，向銀行申請貼現時，除非出口商為該行的授信客戶，否則銀行亦因不明進口商之信用，而為取得債權之保障，須出口商繳納保證金或另提供擔保品始受理貼現。(3)若將跟單匯票委由銀行向進口商託收，雖然是比較穩當的方式，但是貨款之收回需費一段時日，必然影響出口商的資金運用。

有了信用狀，以上兩個問題似乎可以迎刃而解。藉銀行之信用，出口商只要提示符合信用狀條款之匯票及單證，付款自有保障，不必再為資金之運用而煩惱。銀行亦可斟酌開狀銀行之信用而接受押匯。至於進口商方面，亦因開發信用狀，不但不必先付全部貨款，而且既然給與出口商融資之便，故得以向出口商洽議最合算之價格進口貨品。

信用狀的體裁可大別為書信體裁(Letter Style)與統一格式(Uniform Style)。典型的信用狀採書信體裁；蓋信用狀乃開狀銀行的承諾書信，既為書信，必然具備書信的基本要素：信頭〔例8－1，①〕、日期〔例8－1，④〕、受信人〔例8－1，⑤〕、稱呼〔例8－1，⑧〕、本文〔例8－1，⑨~㉕〕、祝頌語〔例8－1，㉗〕及簽署〔例8－1，㉘〕，但識別字首則或有或無。

統一格式的信用狀係指國際商會制定的信用狀格式。國際商會為便利國際間的貿易，配合電腦作業的需要，並顧及未來發展的趨勢，多年來即着手信用狀統一慣例之修訂的同時，分別設計數套信用狀格式，希望藉此統一信用狀內容的編排。

統一格式基本上有五大段落；第一大段保持信頭並標出「不可撤銷信用狀」字樣之外，分左右兩欄〔例8－2，Ⅰ〕；第二大段相當於本文，信用狀的付款條件詳列於此欄〔例8－2，Ⅱ〕；第三大段記載裝運指示〔例8－2，Ⅲ〕；第四大段規定特別條款〔例8－2，Ⅳ〕及第五大段亦分左右兩欄〔例8－2，Ⅴ〕。詳細的內容參閱本章第三節。

二、書信體裁的信用狀

不論採取何種體裁，信用狀的內容必須完整明確而為有關當事人所共同認識者。依作者在外匯部門三十多年之經驗，信用狀的內容可歸納為下

〔例 8－1〕

SHANGHAI COMMERCIAL BANK LTD.
(INCORPORATED IN HONGKONG)
12 QUEEN'S ROAD, CENTRAL
HONG KONG

IRREVOCABLE LETTER OF CREDIT No. 71814
November 29, 1984

K. I Trading Corp.,
No. 30 East Rd.,
Taipei,
Taiwan.

OPENED BY AIRMAIL/CABLE AND ADVISED THROUGH:
Central Trust of China, Banking
& Trust Dept., Taipei, Taiwan.

When opened by cable, this Credit may not be availed of at all unless attached to and made part of our correspondent's notification of such cabled advice, the two jointly constituting evidence of the outstanding amount of this Credit.

Dear Sirs,

At the request of S. Y Company, Wayson Commercial House
Lockhart Road, Wanchai, Hong Kong

we open our Irrevocable Letter of Credit in your favour to the extent of US$6,483.00
Say U.S. Dollars Six Thousand Four Hundred Eighty Three Only
negotiation of

available by your draft(s) in duplicate at --- sight drawn without recourse on
the Applicant for full invoice value accompanied by the following documents at least in duplicate (unless
otherwise specified):-

1. Full set of clean on board bills of lading made out to our order showing the Applicant as notify party
marked "freight prepaid ".

2. Signed commercial invoices in triplicate.

3. Insurance policies or certificates, in assignable form and endorsed in blank for about 10% above invoice
value with claims payable in Hongkong in currency of draft covering Institute Cargo Clauses (W.A.)
(All Risks) and Institute War Clauses, & S.R.C.C.

4. Packing list.

evidencing shipment of 180 Reams Cellophane PT-300, Size: 90cm x 100cm, Weight
per ream: 30 lb., Taiwan origin, Packing: 5 Reams per wooden case, Label: Affix
label on each ream, of which:
150 reams Colour: Colourless, @US$34.30 per ream,
20 reams Colour: Yellow, @US$44.60 per ream,
10 reams Colour: Red, @US$44.60 per ream,
CIF Hong Kong, as per Contract No. O-1036.

This L/C is transferable. In case the drawer presents documents as a trnasferee,
the negotiating bank must certify that the drawer is the transferee of this L/C.

Shipment from Taiwan Port to Hong Kong.
Partial shipments are permitted/prohibited.
Transhipment is permitted/prohibited. Through Bills of Lading are not acceptable.
Shipment latest December 20, 1984.
All documents must be presented for negotiation within 10 days from date of shipment.
This Credit is valid until December 30, 1984 in Taiwan.
The amount of any draft drawn under this credit is to be endorsed on the reverse hereof by the
negotiating bank. All drafts must be marked: "Drawn under Shanghai Commercial Bank Ltd., Hongkong
L/C No. 71814, dated November 29, 1984 ".
We hereby agree with the drawers, endorsers and bona fide holders of all drafts drawn under and in
compliance with the terms of this credit that such drafts shall be duly honoured on due presentation
and delivery of documents as specified.

All banking charges outside Hongkong are for account of beneficiary.

This Letter of Credit is subject to Uniform
Customs and Practice for Documentary
Credits(1983 Revision), International Chamber
of Commerce Publication No. 400.

Yours faithfully,
For SHANGHAI COMMERCIAL BANK LTD.

Fredrick Y. CHU (E-125)   Authorized Signatures

L69 B
100p50x7(V1)

Directions to the negotiating bank are shown on the reverse of this credit.

第八章 交易的履行

· 129

列十大項：

    (1)信用狀號碼及日期，

    (2)信用狀的種類，

    (3)信用狀的當事人，

    (4)信用狀金額，

    (5)匯票，

    (6)必要單據，

    (7)貨品，

    (8)裝運指示，

    (9)信用狀有效日期，

    (10)其他各項。

書信體裁的信用狀大致依照上述次序記載其內容。茲以香港的上海商業銀行第 71814 號信用狀〔例 8−1〕為例(以下簡稱本案)，逐項說明如後：

(一)信用狀號碼及日期(L / C Number and Date)

信用狀由開狀銀行編號，作為日後有關當事人照會之用。本案之號碼為 71814〔例 8−1③〕，又轉知銀行亦有通知號碼可供受益人參照，本案由中央信託局轉知〔例 8−1⑥〕，其號碼為 IAL S06−84−02〔例 8−1⑱〕。信用狀的日期特稱為開狀日期(Issuing Date)。本案的開狀日期為一九八四年十一月二十九日〔例 8−1④〕。屬於信開(By Airmail)的信用狀〔例 8−1⑥〕。

(二)信用狀的種類(Kinds of Credit)

信用狀有六種基本的分類：

1.信用狀用以償付貨款者屬於商業信用狀(Commercial Letter of Credit) 否則為光票信用狀(Clean Letter of Credit)；旅行信用狀(Traveller's Letter of Credit) 則屬於光票信用狀。本案為採購玻璃紙

(Cellophane)而開〔例8－1 ⑮〕，故屬於商業信用狀。

2.信用狀憑匯票附帶必要單據(Required Documents)為付款條件者，稱為跟單信用狀(Documentary Credit)；否則為光票信用狀。本案屬於跟單信用狀〔例8－1 ⑭〕。

3.信用狀的內容得隨時修改或撤銷而毋須預先通知受益人者，稱為可撤銷信用狀(Revocable Letter of Crecdit)；反之未經信用狀有關當事人之同意，不得修改亦不得撤銷者，稱為不可撤銷信用狀(Irrevocable Letter of Credit)。本案為不可撤銷信用狀〔例8－1 ②〕。

4.信用狀經由轉知銀行或第三家銀行承擔與開狀銀行同擔付款義務者，稱為保兌信用狀(Confirmed L／C)；反之則稱為未保兌信用狀(Unconfirmed L／C)。本案為未經保兌的信用狀。

5.信用狀的匯票期限(Tenor)為即期者，稱為即期信用狀(Sight Credit)；反之則為遠期信用狀(Usance Credit)。本案為即期信用狀〔例8－1 ⑬ at……sight〕。

6.信用狀的一部分或全部得由一個或一個以上之其他受益人使用的信用狀，稱為可轉讓信用狀(Transferable L／C)；否則為不可轉讓信用狀(Non－Transferable L／C)。本案為可轉讓信用狀〔例8－1 ⑯〕。以上六種基本的信用狀種類，除「不可撤銷」或「可撤銷」必須標明之外，其餘在信用狀上未特別聲明時，通常可視為商業、跟單、未保兌、即期，而且是不可轉讓的信用狀。至於其他種類的信用狀常見的有：

7.一般信用狀與限押信用狀(General & Restricted Credits)：未指定押匯銀行的信用狀稱為一般信用狀，反之為限押信用狀。限押信用狀常見的句子為 Negotiations under this credit are restricted to … Bank，通常的信用狀並不指定押匯銀行，本案即是。

8.讓購信用狀與直接信用狀(Negotiation & Straight Credits)：開狀銀行對於出票人、背書人及善意持票人均承諾其將兌付符合信用狀條款的跟單匯票時，此信用狀特稱為讓購信用狀。反之僅對出票人承諾者稱為直接信用狀。本案為讓購信用狀〔例8－1 ㉔〕。

9.有追索權信用狀與無追索權信用狀 (With recourse & Without recourse Credits)：信用狀上載有 With recourse 字樣者，謂之有追索權信用狀，反之載有 Without recourse 則爲無追索權信用狀。本案爲無追索權信用狀〔例 8−1 ⑬〕。

10.可回復信用狀與不可回復信用狀 (Revolving & Non−revolving Credits)：信用狀規定在一定期間、一定限額內得以回復使用者，稱爲可回復信用狀，又稱循環信用狀。通常的信用狀爲不可回復信用狀，本案也是不可回復信用狀。

11.本地信用狀 (Local Credit)：信用狀的受益人有時並非供應商，爲實際需要，憑已收到的信用狀向轉知銀行申請另開一信用狀，以供應商爲受益人，此另開的信用狀稱爲本地信用狀。原來的信用狀則稱之爲原主信用狀 (Master L/C)。本地信用狀因以國外開來的信用狀爲後盾，故在美國又稱爲對開信用狀 (Back to Back Credit)。

信用狀尚有其他分類，詳情請參考張錦源編著「信用狀與貿易糾紛」。

(三)信用狀的當事人 (Parties to the L/C)

信用狀有四個基本當事人，即申請人、開狀銀行、轉知銀行及受益人。

1.申請人 (Applicant)：信用狀通常係由進口商按照買賣契約或約定的付款條件，向其往來銀行申請開發，故又稱開發者 (Opener)。書信體裁的信用狀常以 for account of... 載明申請人，因此又稱信用狀的付款人 (accountee)。本案以 at the request of... 表明申請人爲 S Y Company, Hongkong〔例 8−1 ⑨〕。申請人對於開狀銀行爲依照信用狀條款而作付款，負有清償之義務，對外國法律習慣所加於銀行的責任和義務負補償之責。

2.開狀銀行 (Issuing Bank)：循申請人之要求開發信用狀的銀行，稱爲開狀銀行。開狀銀行的名稱地址詳列於信頭。開狀銀行照申請人之指示，儘速將開發信用狀事宜傳達受益人。並以被授權者的立場，接收符合信用狀條款的單據，並對該付款、承兌或讓購銀行履行清償之義務。由於開狀銀行在信用狀交易中扮演最重要的角色，故宜擇國際上具有信譽，富於外匯業務知識與經驗，而且分支機構或往來銀行遍及海外的銀行爲開狀銀行。本案的開狀銀行爲香港的上海商業銀行 (Shanghai Commercial Bank Ltd., Hongkong) 〔例 8–1 ① 〕。

3.轉知銀行 (Advising Bank)：將國外往來銀行開來的信用狀通知受益人的銀行稱爲轉知銀行。轉知銀行與受益人之間並無權利義務或債權債務之關係存在。本案的轉知銀行爲中央信託局〔例 8–1 ⑥ 〕。

4.受益人 (Beneficiary)：信用狀通常以出口商爲受益人。書信體裁的信用狀，其受信人 (Addressee) 即爲受益人或在信用狀內以 in your favor 或 in favor of... 載明受益人。受益人接到信用狀之後，按照規定裝運交貨，即可開出匯票連同必要單據持往押匯，故又稱出票人 (Drawer)。受益人雖然因信用狀而最受惠，但是有遵守信用狀條款的義務，因此應深入了解信用狀條款並加以嚴格解釋。本案的受益人爲 K I Trading Corp., Taipei〔例 8–1 ⑤⑪ 〕。

信用狀除以上四個基本當事人之外，尙有下列當事人。

5.保兌銀行 (Confirming Bank)：若信用狀爲保兌信用狀，則承擔保兌的銀行稱爲保兌銀行。保兌銀行與開狀銀行同擔兌付之責任。本案未經保兌故無保兌銀行。

6.押匯銀行 (Negotiating Bank)：循受益人之請求、承購或貼現信用狀項下之匯票的銀行稱爲押匯銀行。信用狀如無特別指定押匯銀行，則受益人得任意選擇其往來銀行爲押匯銀行(參閱信用狀的種類(7))。

7.付款銀行 (Paying Bank)：信用狀項下匯票的付款人 (Drawee) 爲銀行時(可能是開狀銀行，也可能是第三家銀行)稱爲付款銀行。

8.補償銀行 (Reimbursing Bank)：信用狀規定押匯銀行於押匯後另掣匯票以另一家銀行爲被出票人，並向其求償時，此一被出票人稱爲補償銀行。本案尙有其他對押匯銀行的指示欄印在背面〔例 8–1 ㉙ 〕但未列補償銀行。

㈣信用狀金額（Credit Amount）

信用狀的金額又稱可用金額（available amount），有關金額常見的詞句有：up to an aggregate amount of...；up to an amount not exceeding...；for a sum or sums not exceeding a total of... 等。本案的信用狀金額為陸仟肆佰捌拾叁美元〔例 8－1 ⑫ 〕。

㈤匯票（Drafts）

信用狀必須表明其使用方式為即期付款（by sight payment）、延期付款（by deferred payment）、承兌（by acceptance）或讓購（by negotiation）。本案為讓購方式〔例 8－1 ⑬〕，而以匯票為付款之要件時，則有下列各項記載：

1.出票人（Drawer）：信用狀上用 available by…of your draft(s)的字眼表示出票人。本案的出票人為受益人本身〔例 8－1 ⑬〕。

2.匯票期限（Tenor）：信用狀用 at...sight 表示匯票期限。即期時以 at xx sight 或 at...sight 表示，遠期則將約定的天數打在 at 與 sight 之間，如 at 30 days after sight 或 at 30 days sight 表示之。信用狀交易的匯票期限習慣上為三十天、六十天、九十天、一二〇天、一五〇天至一八〇天。本案的匯票期限為即期〔例 8－1 ⑬〕。

3.被出票人（Drawee）：被出票人即匯票的付款人。信用狀所用的詞句為 drawn on...（被出票人或為進口商或為銀行），本案的被出票人為信用狀的申請人〔例 8－1 ⑬〕。

4.匯票金額（Draft Amount）：匯票的金額與信用狀金額不一定相同。匯票的金額係與交貨的金額成比例，而交貨金額應列明於發票，故信用狀上常用 for.....% invoice value 表示匯票金額。本案規定照發票金額全數出票（for full invoice value）〔例 8－1 ⑬〕。

5.出票條款（Drawn Clause）：信用狀項下的匯票因有跟單而與一般的光票不同。為識別起見，開狀銀行必然要求出票人在匯票上註明信用狀號碼、信用狀日期及開狀銀行。本案的出票條款為 All drafts must be marked " Drawn under Shanghai Commercial Bank Ltd., Hongkong L / C No. 71814 dated November 29, 1984 " 〔例 8 −1 ㉓〕。

(六)必要單據(Required Documents)

信用狀項下應提示的單據因附屬於匯票,故又稱跟單 (accompanied documents),信用狀上以 accompanied by the following documents: 字樣訂明應提示的單據。基本上需要的單據有:

1.商業發票(Commercial Invoice):信用狀項下的貨物以商業發票所載為其全貌,故商業發票具有貨物清單及帳單的性質。商業發票應以信用狀申請人為抬頭人。本案要求一試三份簽署的商業發票〔例 8 −1 ⑭ 2〕。

2.海運 / 海洋提單(Marine / Ocean Bills of Lading):海運提單是信用狀項下最基本的運送單據(Transport Documents),係運送人(Common Carrier)發給託運人(Shipper)有關託運貨物由一地運至另一地的憑證。通常要求全套(Full set)清潔(clean)已裝(on board)的海運提單。所謂全套乃整套三份或兩份而每份都是正本。清潔提單則指提單上無任何加註(Remarks)表示裝載情況在外表上有瑕疵者。海運公司所簽發的提單分已裝提單(Shipped B / L)與備裝提單(Received B / L);前者憑該公司的簽署,後者則加簽 On Board 字樣並簽署表示已裝。至於提單的受貨人(Consignee)則要求可轉讓的指示式提單(Order B / L)為多。本案即要求以開狀銀行為抬頭的指示式提單,並註明以進口商為到貨通知單位(Notify Party)及運費付訖(Freight Prepaid)等字樣〔例 8 −1 ⑭ 1〕。

3.保險單據(Insurance Documents):信用狀要求的最低保險金額為商業發票金額加百分之十。對於運輸契約的全航程之危險付保,並投保該交易通常應保之險類、條款。本案要求保單或保險證明書(Insurance policies or certificates),投保全險(All Risks)加保兵險及罷工暴動險(Institute War Clauses, & S. R. C. C.)〔例 8 −1 ⑭ 3〕。

4.裝箱單(Packing List):裝箱單也是必要單據中常見者,除散裝貨

(Bulk Cargo)以外，幾乎不可或缺的單據，為檢驗、驗關、提貨時核對項目之用。本案亦要求裝箱單〔例 8－1 ⑭ 4 〕。以下的單據則屬於次要但常見者：

5.海關發票(Customs Invoice)：外銷美國、加拿大、澳洲及紐西蘭等地須提出各該國所規定之進口報關所需發票。

6.產地證明書(Certificate of Origin)：為貨品確屬某地製造或生產之憑證。

7.重量尺碼證明書(Weight / Measurement Certificate)：以重量買賣或尺碼計價的貨品需提供此項證明。

8.檢驗證明書(Inspection Certificate)：為防止製造不合標準品質或合約規格的貨品，或配合輸入國海關之需要而出具的證明文件。由製造廠商出具者稱為 Manufacturer's or Maker's Inspection Certificate，而由公證行出具者稱為 Independent Inspection Certificate。

㈦貨品(Commodities)

貨品為信用狀付款的標的物，應記於各項單據上。信用狀所用的詞句為 evidencing shipment of … 。有關貨品的敘述主要的是數量(Quantity)、名稱規格(Description)及貿易條件(Trade Terms)等。數量的表示法若有(about)字樣者裝載時允許百分之十上下的差數，而無此規定時僅允許百分之五上下之範圍但以不超過信用狀金額為限，惟包裝單位或個件為數量單位者不適用。各項單據中，商業發票上的貨物規格應與信用狀所載的規格相同。其餘的單據上，貨物得填列與信用狀貨物規格並無抵觸之統稱(General Terms)。本案的貨品為一百八十令的玻璃紙，規格細節則以合約為準(as per Contract No. 0－1036)〔例 8－1 ⑮ 〕。

㈧有關裝運指示(Shipping Instructions)

信用狀上通常載明的裝運指示有：

1.裝貨港及卸貨裝(Loading & Discharging Ports)：本案自台灣的港

口起運至香港〔例 8－1 ⑲ 〕。

2.裝運期限(Latest Shipment Date)：有關最遲裝運日期應確切表示，至於裝運日期則以提單的已裝日期(on board date)爲準。本案的裝運期限爲「一九八四年十二月二十日」〔例 8－1 ⑲ 〕。

3.分批裝運(Partial Shipments)：除非信用狀另有規定，分批裝運是可以的。本案明確表示可以分批裝運〔例 8－1 ⑲ 〕。

4.轉運(Transhipment)：將裝載於船上的貨物卸下，再裝載於另一船舶，稱爲轉運或轉船。本案明確地表示不可轉運〔例 8－1 ⑲ 〕。

5.艙面裝運(On Deck Shipment)：一般的貨品(General Merchandise)通常不得裝在艙面，因此准許艙面裝運的貨品，信用狀應加以註明。

(九)信用狀有效日期(Expiration)

信用狀不論其爲可撤銷或不可撤銷，必須規定提示單據之有效日期以便付款、承兌或讓購。除了有效日期之外，信用狀尚須載明裝運日後應提示單據之特定期間，作爲提示單據的期間。本案的有效日期爲一九八四年十二月三十日〔例 8－1 ㉑ 〕，而單據的提示期間爲裝船後十天〔例 8－1 ⑳ 〕。

(十)其他(Others)

信用狀除了以上的記載、規定之外，尚有一項附加的條款和三項印好的固定條款如下：

1.特別條款(Special Instructions)：非經常性的指示或條款應擇空欄明確記載。本案的手續費條款〔例 8－1 ㉕ 〕係屬特別條款。上述裝運日後提示單據之特定期間〔例 8－1⑳ 〕亦可視爲特別條款之一。

2.對押匯銀行的指示(Instructions to Negotiating Bank)：開狀銀行爲便於信用狀項下交易能順利，對押匯銀行常有以下的要求：(1)押匯金額的背書指示(Endorsement Instructions)；(2)寄單指示(Mailing Instructions)及(3)補償指示(Reimbursement Instructions)等。本案對押匯銀行背書要求載於正面〔例8－1 ㉑〕，而另於背面要求郵寄單證及求償方式〔例8－1 ㉙〕。

3.開狀銀行的承諾條款（Agreement Clause）：信用狀為開狀銀行對於符合信用狀條款的跟單匯票，履行付款或承兌或讓購的承諾書信，因此對於有關當事人之承諾應有所載明。本案係對於出票人、背書人以及善意持票人的承諾〔例8－1 ㉔〕，故而為讓購信用狀〔參閱信用狀的種類(8)〕。

4.有關規定（Provisions）：為期消除國際貿易的諸種障礙，尤其是為達到各國或各地區對於信用狀的定義，內容有共同的認識與解釋，國際商會（International Chamber of Commerce）於一九三三年第七次總會時制定商業信用狀統一慣例（Uniform Customs & Practice for Commercial Documentary Credits）呼籲世界各國採納。該項慣例於一九五一年里斯本第十三次總會時第一次修訂。第二次修訂案於一九六二年十一月提出，而為一九六三年墨西哥第十九次總會確認並自一九六三年七月一日起實施，從此稱為信用狀統一慣例（Uniform Customs & Practice for Documentary Credits 1962 Revision, ICC Brochure No. 222）。但是世界貿易區域之擴張，交易之劇增，使運輸方式起了極大的改變，如貨櫃運輸、聯合運輸之興起，在在使上項慣例之諸多條款無法適應目前貿易之主流及未來發展的趨勢；於是一九七四年十二月第三次修訂後於一九八三年再度修訂了統一慣例〔Uniform Customs & Practice for Documentary Credits (1983 Revision) International Chamber of commerce Publication No. 400〕。並自一九八四年十月一日起實施。時隔不及十年，又為因應實務上的需要，經一九九三年四月國際商會執行委員會通過而於同年五月發行修訂本，編列為國際商會500號出版物（Uniform Customs & Practice for Documentary Credits 1993 Revision, ICC Publication No. 500）。本案則適用舊的400號版本〔例8－1 ㉖〕。

本案信用狀內容尚有四項未歸類者分述如下：

⑦本條款乃上海商業銀行印便之條款，係配合信用狀統一慣例之規定，以電報或電報交換開發信用狀並以「郵寄電報證實書」為有

效之信用狀時，須於電報上載明應俟收到郵寄之證實書後始生效力。故該行的信用狀若以電報開發者，特別要求連同轉知銀行的電報通知書一併提示始能生效。本案由⑥可知係航郵開發(Opened by Airmail)的信用狀，故不適用⑦項條件辦理。

⑧乃稱呼，僅屬於書信必備的要項，不列入信用狀的條款。

㉗乃結尾客套，亦屬於書信必備的要項，不列入信用狀的條款。

㉘乃簽署，亦屬於書信必備的要項，但因係負文責，與承諾條款前後呼應，而歸入該項條款為宜。

三、統一格式的信用狀

如前所述統一格式的信用狀其內容係配合電腦繕打處理而事先予以編排各項內容的位置，可大別為五大段，茲以新加坡的馬來亞銀行(Malayan Banking Berhad, Singapore)開來的 S／94／49 號信用狀〔例 8－2〕為例(以下簡稱乙案)說明如後：乙案的信頭除開狀銀行的簡稱(Maybank)及標誌老虎頭〔例 8－2 ①〕之外，表示開狀地點為新加坡〔例 8－2②〕，以快遞郵寄方式〔例 8－2③〕於一九九四年七月一日開狀〔例 8－2④〕。

第一大段：在信頭之下分左右兩欄。左欄分三項：由上而下為信用狀種類、轉知銀行及受益人。乙案為不可撤銷的跟單信用狀〔例 8－2⑤〕，經由第一波士頓銀行台北分行轉知〔例 8－2⑥〕給受益人，台北的 NPC 公司〔例 8－2⑦〕。右欄則分四項：由上而下依序為信用狀及轉知編號、申請人、信用狀金額及有效日期。乙案的信用狀號碼為 S／94／49〔例 8－2⑧〕，轉知銀行的編號為 BOB－4097〔例 8－2⑨⑰〕。申請人為新加坡的 SPE 公司〔例 8－2⑩〕。信用狀金額為美金 13,950 元(美金壹萬叁仟玖佰伍拾元正)〔例 8－2⑪〕。有效日期為受益人所在地時間，即台灣當地時間一九九四年十月八日〔例 8－2⑫〕。

第二大段：相當於本文部份。主要的記載事項有：匯票條款、必要單據及貨品。乙案的第二大段以稱呼〔例 8－2 ⑬〕開頭，開狀銀行對受益人聲明開發不可撤銷的跟單信用狀〔例 8－2 ⑭〕。其付款條件係憑受益人出具一式兩份即期、以開狀銀行為付款人(被出票人)、照發票全額開立的匯票〔例 8－2 ⑮〕。

必須提示的跟單〔例 8－2 ⑯〕有三：(1)一式四份的經過簽署的商業發票〔例 8－2(16－1)〕；(2)全套清潔已裝、以開狀銀行為受貨人、到埠通知申請人並註明運費預付的海運提單〔例 8－2(16－2)〕；及(3)經空白背書投保發票金額加一成，保險種類含協會貨物 A 條款、兵險及罷工險並理賠地為新加坡的保險單據〔例 8－2(16－3)〕。

乙案的貨物以通稱扼要列述為 9,000 公斤的 NP 纖維〔例 8－2 ⑱〕，由於乙案係以重量為買賣單位，故得有百分之十的寬容，連帶地金額亦得有相同的寬容範圍〔例 8－2 ⑲〕。

第三大段為有關裝運指示。乙案則再分隔為上下兩欄，上欄為裝貨港及卸貨港。乙案由台灣運至新加坡〔例 8－2 ⑳〕。下欄由左而右為最遲裝運日期、分批及轉運事項。乙案規定最遲應於一九九四年九月三十日之前裝運〔例 8－2 ㉑〕，不可分批〔例 8－2 ㉒〕，亦不可轉運〔例 8－2 ㉓〕。

第四大段為特別條款欄。乙案有三項指示：(1)新加坡地區以外之銀行費用由受益人負擔〔例 8－2 ㉔〕；(2)提示單據之特定期間限於裝載日之後八天內惟不得逾越信用狀有效日期〔例 8－2 ㉕〕；(3)匯票必須註明出票條款〔例 8－2 ㉖〕，此項出票條款其實非為特別條款而係一般的匯票條款。

第五大段分左右兩欄：左欄為開狀銀行的承諾條款，右欄為對押匯銀行的指示。乙案的開狀銀行在左欄承諾〔例 8－2 ㉗〕，並於祝頌語〔例 8－2 ㉘〕之後加以簽署〔例 8－2 ㉙〕。在右欄的對押匯銀行的指示有三〔例 8－2 ㉚〕：(1)押匯銀行必須於信用狀背面背書；(2)押匯的匯票及單據應分兩梯次以掛號郵寄至開狀銀行及(3)請向台北的第一波士頓銀行提示匯票及單據以便補償。

① **Maybank**

ORIGINAL

② Place : Singapore

③ ☒ Operative credit instrument forwarded to advising bank by xxxxx courier service
☐ Operative credit instrument confirming our pre-advice by cable/telex of today's date.

④ Date of Issue:　1 Jul 94

⑤ IRREVOCABLE DOCUMENTARY CREDIT	ISSUING BANK'S NO.	ADVISING BANK'S NO.
	⑧　S/94/49	⑨

Advising Bank
⑥ THE FIRST NATIONAL BANK OF BOSTON
United Commercial Building, 5th Fl
137 Nanking East Road
Section 2, Taipei 10409
Taiwan

⑩　S　　P　　E
Applicant

Singapore

Beneficiary
⑦ M/s N　　P　　C

Taipei
Taiwan

Amount
⑪ USD13,950-00 CIF
United States Dollars: Thirteen thousand
nine hundred and fifty only.

Expiry date for negotiation in country of beneficiary
⑫　8 Oct 94

⑬ Dear Sirs,

⑭ We hereby issue in your favour this irrevocable documentary credit ⑮ which is available by your drafts in
duplicate at ⑯ xxxx sight drawn on Issuing bank.

for invoice value accompanied by the following documents (in duplicate unless otherwise specified):-

⑯-1 Signed Commercial Invoices in quadruplicate.

⑯-2 Full set of clean "on board" Marine Bills of Lading made out to the order of Malayan Banking Berhad, notify
applicant and marked "freight prepaid".

⑯-3 Insurance policies (or certificates) endorsed in blank, for invoice value of the goods plus 10% covering Marine
and War Risks, including Institute Cargo Clauses A　　　. Institute War Clauses (Cargo) and Institute
Strikes Clauses (Cargo)

Claims payable at　　　Singapore
xxxxxxxxxxx　　　　　　　　　　xxxxxx

⑰ 信　用　狀　通　知　章
通知編號 BOB- 4097
日　期　JUL - 4 1994
THE FIRST NATIONAL BANK
OF BOSTON TAIPEI BRANCH

⑱ Covering:　9,000 kg N.　.P　　Fibre 7D x 64mm K-Type Khaki
Hollow/Conjugate/silicone (1 FCL)
Packing in bale.

⑲ Beneficiary may draw 10% more or less on amount and quantity.

⑳ SHIPMENT FROM
Taiwan
TO
Singapore

㉑ LATEST SHIPMENT DATE	PARTIAL SHIPMENT	TRANSHIPMENT
㉑ 30 Sep 94	㉒　Prohibited	㉓　Prohibited

㉔ SPECIAL CONDITIONS:
All bank charges including reimbursement charges outside Singapore are for beneficiary's account. ㉕ The documents must be presented not later than
8　　　days after the date of issuance of the bills of lading (or other shipping documents). Nevertheless such presentation must
not be after the expiry of the credit. ㉖ Drafts drawn under this credit must bear the following clause "Drawn under Malayan Banking Berhad, Singapore
credit number　S/94/49　　dated　　1 Jul 94

㉗ We hereby engage with drawers and/or bonafide holders that drafts drawn and
negotiated in conformity with the terms of this credit will be duly honoured on
presentation and that drafts accepted within the terms of this credit will be duly
honoured at maturity.

㉘ Yours faithfully,
Maybank

㉙
Authorised Signatures

Malayan Banking Berhad
Form IE 3 (1/76)

㉚ INSTRUCTIONS TO THE NEGOTIATING BANK
The amount of each drawing must be endorsed on the reverse of this credit. All
drafts and documents are to be forwarded to us in two sets by consecutive
registered airmail.
REIMBURSEMENT INSTRUCTIONS:

Please present documents and drafts
to the First National Bank of Boston,
Taipei for reimbursement and
disposal.

THIS CREDIT IS SUBJECT TO THE UNIFORM CUSTOMS AND PRACTICE FOR DOCUMENTARY CREDITS, INTERNATIONAL CHAMBER OF COMMERCE PUBLICATION NO 400

Advice for the beneficiary

乙案的有關約定則繕打於左邊欄外，表明本信用狀遵照國際商會第 500 號版本的信用狀統一慣例。以上各大段的內容亦可參照第二節的分類法歸納為十大項(參閱練習問題三)。

四、SWIFT 方式的信用狀

　　以電報開發的信用狀，可分為 TELEX 拍發者和 SWIFT 拍發者。前者因易受干擾有被竄改之虞。雖然電文上加密碼(Test Key)以供識別真偽，但需經人工核算和解密，費時費力。自一九七三年由十五個國家的 239 家銀行在比利時的首都布魯塞爾成立環球銀行財務電信協會(Society for Worldwide Interbank Financial Telecommunication 簡稱 SWIFT)以來，因經由 SWIFT 處理銀行業務，不但傳送迅速，格式統一，易於追蹤控制，最主要者乃其安全性極高。我國的外匯銀行亦自一九八五年起陸續加入為會員，因此愈來愈多的信用狀係經由 SWIFT 拍發。

　　茲以日本的京都銀行(The Bank of Kyoto, Ltd., Kyoto)開來的第 31 − 0274 號信用狀為例(以下簡稱丙案)說明如下。

　　丙案係經由中國國際商業銀行(The International Commercial Bank of China, Taipei)轉知的格式。故信頭為轉知銀行〔例 8 −3 之一①〕並標明為跟單信用狀的通知書〔例 8 −3 之一②〕，通知受益人，台北的 FHF 公司〔例 8 −3 之一③〕。通知日期為一九九六年十月八日，轉知編號為 AHL1813〔例 8 −3 之一④〕。在稱呼〔例 8 −3 之一⑤〕之後轉知銀行聲明，在不受約束之情況下，將收自京都銀行經確認真實性的 SWIFT 電文照轉如下〔例 8 −3 之一⑥〕。此際 BOKFJPJZ 係京都銀行的代碼。來電的內容自 Quote 起至 Unquote 為止。我們要注意的是來電的真實性(Authenticity)而其真實性在 SWIFT 電文中會自動顯示，即由〔例 8 −3 之一⑦〕得知丙案經確認其真實性(Auth Ok)。並從〔例 8 −3 之一⑧〕亦知本案採 MT700 係屬於詳電，故為有效的信用狀(Operative Credit)，無需再寄發郵寄確認書(Mail Confirmation)。MT700 / 701〔例 8 −3 之一

⑨〕表示電文包括兩頁。

　　丙案的電文均附有欄位號數及其內容，依序爲：

27：序列總數 1／2 表示兩頁的電文中之第一頁

40A：信用狀的種類爲不可撤銷

20：跟單信用狀的號碼爲 31−0274

31C：開狀日期爲一九九六年十月七日

31D：有效日期爲台灣時間一九九六年十一月十五日

50：申請人爲 MAS CO.,LTD., KYOTO

59：受益人爲 FHF， TAIPEI

32B：信用狀金額爲美金 47, 880. 00

41D：付款方式可透過任意銀行以押匯方式辦理

42C：開立即期照發票金額的匯票

42A：付款人(被出票人)爲京都銀行(BOKFJPJZ)

43P：可以分批裝運

43T：不可轉運

44A：自台灣港口裝運

44B：裝運至大阪

　　以下電文內容參閱〔例 8−3 之二〕

44C：最遲裝運日期爲一九九六年十月三十一日

45A：貨品爲 19,200 磅的大華牌混紡紗，每磅美金 1.50 詳如第 520 號合
　　約，及 18,000 磅的大華牌粗紡，每磅美金 1.06 詳如第 209 號合
　　約，貿易條件爲 CIF。

71B：日本地區以外之銀行費用由受益人負擔

48：提示單據之特定期間限於裝運日之後起算 15 天內，但不得逾越有效
　　日期。

49：不需附加保兌

53A：補償銀行爲 BOTKUS33 (參考 78)

78：對押匯銀行之指示有三：(1)指定紐約的東京三菱銀行爲補償銀行；

(2)匯票及單據應一次寄送及(3)補償事宜遵照國際商會 525 號版本的補償
　　規則。

72：本信用狀遵照國際商會第 500 號版本的信用狀統一慣例

　　以下為電文的第二頁〔例 8–3 之三〕

27：序列總數 2/2 表示兩頁的電文中的第二頁

20：跟單信用狀號碼係 31 –0274 號

46B：必要單據

　　1)一式三份已簽署的商業發票

　　2)全套(減一份)清潔已裝、以申請人為受貨人、註明運費預付、到
　　　埠通知申請人的海洋提單。

　　3)空白背書，投保金額為發票金額加一成，含全險、兵險及罷工暴
　　　動險，投保幣別與匯票之幣別相同，並在日本理賠的海上保險單
　　　據。

　　4)一式三份的裝箱單

　　5)受益人的證明書敘明一套副本單據，包括 1/3 正本提單及乙份正
　　　本的優惠關稅產地證明書，業已以國際快遞郵件(EMS)寄送申請
　　　人。

47B 特別條款：不得以電匯方式補償

UNQUOTE 表示來電內容結束

　　轉知銀行特地提醒受益人本案的 SWIFT 信用狀係遵照國際商會第
500 號版本的信用狀統一慣例。

　　丙案亦得比照第二節的分類法歸納為十大項(參考練習問題四)。

五、開發信用狀

(一)進口簽證

　　我國的進口廠商欲進口管制類或需簽證之准許類貨品，須分別向國際

〔例8－3之一〕

① THE INTERNATIONAL COMMERCIAL BANK OF CHINA
HEAD OFFICE-FOREIGN DEPARTMENT

100 CHI LIN ROAD
TAIPEI 10424 TAIWAN
REPUBLIC OF CHINA
TEL:5633156

ADDRESS: INTCOMBANK
Y ADDRESS: ICBCTWTP007
TELEX 11300 INCOBK 22145 INTCOMBK
FAX 5632614

② Notification of Documentary Credit

Beneficiary ③	Date:	OSN:	Page:1
F:... H:... F 2.., T. H... ROAD TAIPEI, TAIWAN REPUBLIC OF CHINA	④ OCT. 08, 1996 9973 Advising No :AHI. :1813		BANK ID: 250. 列印序號 : :9 頁 次 : 43

⑤ Dear Sirs,
⑥ Without any responsibility and/or engagement on our part, we have the
pleasure of advising you that we have received an authenticated S.W.I.F.T.
/TELEX message from THE BANK OF KYOTO, LTD.   (BOKFJPJZ   )
FOREIGN DEPT.
Reading as follows: QUOTE
****
* LABEL: QSH :2833   N007.   96/10/08
* MSGACK:DWS76 ⑦ Auth OK, key . 960915   ICBCTWTP BOKFJPJZ record
* ⑧ MT 700 DOCUMENTARY CREDITS ⑨ MT 700/701
****
         {1: ICBCTWTP.
         {2: 0017099961007BOKFJPJZ
         {4:
:27 :SEQUENCE OF TOTAL
     1/2
:40A:FORM OF DOC. CREDIT
     IRREVOCABLE
:20 :DOCUMENTARY CREDIT NUMBER
     :31-0274
:31C:DATE OF ISSUE
     961007
:31D:EXPIRY DATE & PLACE
     961115TAIWAN
:50 :APPLICANT
     MAS CO., LTD.
     :4, KAMI   YANAGI-CHO, HIRANO
     KITA-KU, KYOTO, JAPAN
:59 :BENEFICIARY
     F: H   F
     2.., T. H   ROAD TAIPEI, TAIWAN
     REPUBLIC OF CHINA
:32B:CURRENCY CODE, AMOUNT
     USD47880,00
:41D:AVAILABLE WITH...BY...
     ANY BANK
     BY NEGOTIATION
:42C:DRAFTS AT ...
     DRAFT AT SIGHT FOR FULL INVOICE
     VALUE
:42A:DRAWEE
     BOKFJPJZ
:43P:PARTIAL SHIPMENTS
     PERMITTED
:43T:TRANSSHIPMENT
     PROHIBITED
:44A:LOADING IN CHARGE AT/FROM
     TAIWANESE PORT
:44B:FOR TRANSPORTATION TO...

注 意 事 項 *** T(BE CONTINUED ON NEXT PAGE   ***
請詳細校閱本通知書內容,如有不能接受或應
行者,務即直接逕洽買方修改,以利押匯手續。

〔例8－3之二〕

THE INTERNATIONAL COMMERCIAL BANK OF CHINA
HEAD OFFICE-FOREIGN DEPARTMENT

CABLE ADDRESS: INTCOMBANK
SWFT ADDRESS: ICBCTWTP007
TELEX: 11300 INCOBK 22145 INTCOMBK
FAX: 5632814

100 CHI LIN ROAD
TAIPEI 10424 TAIWA
REPUBLIC OF CHINA
TEL:5633156

Notification of Documentary Credit

Beneficiary	Date:	OSN:	
F H F	OCT. 08, 1996 9973		Page:2
2 T H ROAD TAIPEI, TAIWAN			BANK ID: 250
REPUBLIC OF CHINA	Advising No AHL 1813		列印序號 :
			頁 次 : 1

```
               *** CONTINUED ***
          OSAKA
:44C:LATEST DATE OF SHIPMENT
          961031
:45A:SHIPMENT (OF GOODS)
          19,200LBS OF ''TA HWA         BRAND
                              BLENDED YARN, RAW WHITE ON CONE

                   UNIT PRICE USD1.50 PER LB
                   AS PER CONTRACT NO.   520
          18,000LBS OF ''TA HWA      BRAND
                              RAW WHITE, ON CONE
                   UNIT PRICE USD1.06 PER LB
                   AS PER CONTRACT NO.   209
          CIF
:71B:CHARGES
          ALL BANKING CHARGES OUTSIDE JAPAN
          ARE FOR ACCOUNT OF BENEFICIARY
:48 :PERIOD FOR PRESENTATION
          DRAFTS AND DOCUMENTS MUST BE
          PRESENTED WITHIN 15 DAYS AFTER
          SHIPPING DATE BUT WITHIN
          CREDIT VALIDITY
:49 :CONFIRMATION INSTRUCTIONS
          WITHOUT
:53A:REIMBURSEMENT BANK
          BOTKUS33
:78 :INSTR TO THE PAY/ACC/NEG BANK
          PLEASE CLAIM REIMBURSEMENT ON THE BANK OF TOKYO-MITSUBISHI,
          LTD., NEW YORK ATTN: IOD REIMB. SEC.
          NEGOTIATING BANK MUST SEND DRAFTS AND ALL DOCUMENTS TO
          ISSUING BANK IN ONE LOT
          REIMBURSEMENT IS SUBJECT TO ICC URR 525
:72 :BANK TO BANK INFORMATION
          THIS CREDIT IS SUBJECT TO U.C.P.
          1993 I.C.C. PUBLICATION NO.500
          -}
          {5:(MAC: F336 )
          {CHK: E2DO B206}
          }

     UNQUOTE

          *** TO BE CONTINUED ON NEXT PAGE ***
```

注 意 事 項

請校閱本通知書內容，如有不能接受或願
運洽買方修改，以利押匯手續

貿易英文

146

〔例 8－3之三〕

THE INTERNATIONAL COMMERCIAL BANK OF CHINA
HEAD OFFICE-FOREIGN DEPARTMENT

ADDRESS INTCOMBANK
ADDRESS: ICBCTWTP007
TELEX 11300 INCOBK 22145 INTCOMBK
FAX 5832614

100 CHI LIN ROAD
TAIPEI 10424 TAIWAN
REPUBLIC OF CHINA
TEL:5633156

Notification of Documentary Credit

Beneficiary		Date:	OSN:	Page: 3
F :II F.		OCT. 08. 1996	9973	
2 T II ROAD TAIPEI, TAIWAN				BANK ID: 250
REPUBLIC OF CHINA		Advising No AHI. 1813:		列印序號 : 7
				頁 次 : 14

```
                    * * *   CONTINUED   * * *
QUOTE
*****************************************************************************
* LABEL: QSN  2833.  N007   96/10/08
* MSGACK:DWS76.  Auth OK, key   960915         ICBCTWTP BOKFJPJZ record
*          MT 701 DOCUMENTARY CREDITS    MT 700/701
*****************************************************************************
        (1:   ICBCTWTP.
        (2:   011709961007BOKFJPJZ.
        (4:
:27 :SEQUENCE OF TOTAL
      2/2
:20 :DOCUMENTARY CREDIT NUMBER
      :31-0274
:46B:DOCUMENTS REQUIRED
      1)SIGNED COMMERCIAL INVOICE IN TRIPLICATE
      2)FULL SET(LESS ONE) OF CLEAN ON BOARD OCEAN BILLS OF LADING MADE
        OUT TO MAS   CO., LTD. AND MARKED ''FREIGHT PREPAID''
        AND SHOWING THE ABOVE APPLICANT AS ''NOTIFY PARTY''
      3)MARINE INSURANCE POLICY OR CERTIFICATE ENDORSED IN BLANK FOR
        110 PERCENT OF INVOICE COST INCLUDING INSTITUTE CARGO CLAUSES
        (ALL RISKS), INSTITUTE WAR CLAUSES, INSTITUTE S.R.C.C. CLAUSES
        AND CLAIMS TO BE PAYABLE IN CURRENCY OF DRAFTS IN JAPAN
      4)PACKING LIST IN TRIPLICATE
      5)BENEFICIARY'S CERTIFICATE STATING THAT A DUPLICATE SET
        DOCUMENTS INCLUDING 1/3 ORIGINAL B/L AND G.S.P. FORM A IN ONE
        ORIGINAL HAS BEEN DESPATCHED DIRECT TO APPLICANT BY EMS SPEED
        POST
:47B:ADDITIONAL CONDITION
      T.T. REIMBURSEMENT IS PROHIBITED
      -)
      (5:(MAC:  8325  )
      (CHK:  32E3D1F   )
      )

UNQUOTE
        Unless specifically stated, the documentary credit received from
S.W.I.F.T. is subject to the U.C.P. for Documentary Credit, I.C.C. PUB.
NO.500 of 1993, which are in effect on the date of issue.
```

<table>
<tr><td>

注 意 事 項

請詳細校閱本通知書內容，如有不能接受或應
行者，務即直接逕洽賣方修改，以利押滙手續。

</td></tr>
</table>

第八章　交易的履行

貿易局或其委託之簽證銀行申請輸入許可證。目前進口大陸農產品及工業產品係屬於簽證項目，須繕妥輸入許可證申請書〔例 8-4〕，向簽證銀行申請輸入許可證。此項許可證一套三聯，除第一聯為輸入許可證申請書〔例 8-4〕，第二聯為輸入許可證(副本)，第三聯為輸入許可證(正本)〔例 8-5〕。繕打申請書時應注意事項，請參考申請書背面的各欄填寫說明(附錄三)。

(二)申請開發信用狀

進口廠商向指定銀行申請開發信用狀，其貨品屬於免簽證項目者憑交易憑證，如合約、訂單或預估發票等；若其貨品屬於需簽證項目者，則須憑輸入許可證向指定銀行申請開發信用狀。在申請之前宜先考慮結匯價款(俗稱保證金)。申請開發信用狀時將信用狀金額，按當日賣出匯率折合新台幣結繳指定銀行，由其透過外匯市場買入所需外匯是為結購外匯，通稱為結匯。結匯的成數可分為全額結匯和部分或免結匯；前者按信用狀金額全數一次結匯，而後者則先結匯一部分或全數不結匯，日後再分次結匯。全額結匯對於指定銀行而言無任何外匯風險，因此甚少要求申請人另定契約或提供擔保者；部分結匯或免結匯因具有外匯及信用等風險，申請人須先經指定銀行授信後始能辦理開狀。常見的授信方式為訂定進口押匯契約和購料貸款契約。結匯成數因貨品的市場性及廠商的信用狀況而有別，若約定為部分結匯則於開發信用狀時依約定比例先行結匯外，餘額則於約定日期再行結匯，如進口押匯通常是貨到贖單時結匯；而購料貸款則於到期日再行結匯歸還指定銀行所墊款項並繳付利息。結匯成數關係業務資金的運用甚巨，故宜及早接洽指定銀行辦妥有關訂約事項。結匯成數決定之後，即可填寫開發信用狀申請書向指定銀行申請開發信用狀。申請書的格式(參閱本章練習問題八)雖然每家銀行略有不同，但是其內容與前述信用狀內容類同，認識信用狀者不難填寫。惟填寫時宜注意：

1. 信用狀金額應與交易憑證或輸入許可證之金額一致。
2. 貨品名稱不宜太詳細。

〔例 8-4〕

輸 入 許 可 證 申 請 書
APPLICATION FOR IMPORT PERMIT

第一聯：簽證機構存查聯

| 共 | 頁 | 第 | 頁 |

①申請人 Applicant	③生產國別 Country of origin	④起運口岸 Shipping Port
②申請人印鑑 Signature of Applicant	⑤賣方名址 Seller	
	⑥發貨人名址 Shipper	
	⑦檢附文件字號	
（請蓋國際貿易局登記之印鑑）		

⑧項次 Item	⑨貨品名稱、規格、廠牌或廠名等 Description of Commodities Spec. and Brand or Maker, etc.	⑩商品分類號列 及檢查號碼 C.C.C. Code	⑪數量及單位 Q'ty & Unit	⑫單價 Unit Price	⑬條件及金額 Terms & Value

簽證機構加註有關規定 Special Conditions	輸入許可證號碼 Import Permit No.
	許可證簽證日期 Issue Date 許可證有效日期 Expiration Date
	簽證機構簽章 Approving Agency Signature

一、本輸入許可證自一經塗改即為失效，商品分類號列非蓋有該簽證機構校對章者無效。
二、本輸入許可證如有塗改等情事，關係商業機密，請予保密，不得對外洩漏資料。
三、進口貨品，申請人應自行瞭解及依照有關輸入規定，報驗、檢疫、衛生及其他相關國內管理法令辦理。

| 收件號碼 |
| 收件日期 |

〔例 8-5〕

輸入許可證（正本）
IMPORT PERMIT（ORIGINAL）

第三聯：進口人報關用聯

	共　　頁　第　　頁

①進口人 Importer	③生產國別 Country of origin	④起運口岸 Shipping Port
②進口人印鑑 Signature of Importer	⑤賣方名址 Seller	
	⑥發貨人名址 Shipper	
（請蓋國際貿易局登記之印鑑）	⑦檢附文件字號	

⑧項次 Item	⑨貨品名稱、規格、廠牌或廠名等 Description of Commodities Spec. and Brand or Maker, etc.	⑩商品分類號列及檢查號碼 C.C.C. Code	⑪數量及單位 Q'ty & Unit	⑫單價 Unit Price	⑬條件及金額 Terms & Value

簽證機構加註有關規定 Special Conditions	輸入許可證號碼 Import Permit No.
	許可證簽證日期 Issue Date
	許可證有效日期 Expiration Date
	簽證機構簽章 Approving Agency Signature

一、本輸入許可證一經塗改即屬失效，商品分類號列蓋有簽證機構校對章者除外。
二、本輸入許可證記有貿易資料，關係商業機密，請予保密，不得外漏或買賣。
三、進口貨品，申請人應自行瞭解及依照有關輸入規定、檢驗、檢疫、衛生及其他
　　相關國內管理法令辦理。

收件號碼
收件日期

3. 信用狀內容不得與交易憑證或輸入許可證所列內容抵觸。

4. 參照合約訂出適當的最遲裝運日期，但不得逾越輸入許可證有效日期。

5. 宜訂明在裝船後幾日必須提示單據押匯，以免出口商拖延提示單據，而延誤提貨時間遭受損失。

6. 配合注意事項 5，訂出信用狀的有效日期。

7. 依照約定以書信或電報開發信用狀。

信用狀申請書填妥後即行繳納結匯價款和有關費用，其計算式如下：

1. 結匯價款 ＝信用狀金額(外幣)×(結匯成數)×(當日賣出匯率)。

2. 開發信用狀手續費 ＝信用狀金額(外幣)×費率×(當日賣出匯率)。

   費率的計算法以三個月為一期，各期費率及最低收費由各指定銀行自行訂定。

3. 郵費：按投遞之地區分別收費。

詳情請洽各指定銀行。

六、接受和修改信用狀

㈠接受信用狀

出口廠商於接獲國外開來的信用狀，應調閱與該信用狀有關的交易確認函件，例如銷貨確認函、買賣合約或訂貨單等，並參照下列步驟核對信用狀條款以便決定是否接受。蓋信用狀或以交易的合約為基礎，但是信用狀在本質上與買賣合約或其他合約係屬分立的交易行為，而且開狀銀行或押匯銀行均與該等合約無關，也不受該等合約的拘束。由於出口廠商於接受信用狀後須先盡交貨的義務方能享受貨款的權利，因此不能不慎重。核對的方法可以參考下列的步驟：

步驟一：查對信用狀的可靠性

經由轉知銀行正式通知的信用狀原則上屬於真實而非偽造者。若轉知銀行在通知書上書明「無法驗對印鑑」或「無業務往來關係」則該信用狀的真實性無法確定；而電報開發的信用狀，其電報上的押密不符時該信用狀的有效性待查，須俟押密核符後始為有效；所收信用狀必須標明不可撤銷的信用狀，否則其內容有遭片面修改之虞；若開狀銀行的情況或當地情勢不穩定，則信用狀宜經由指定而可靠的一流銀行保兌。

步驟二：核對交貨有關條件

1. 由信用狀的受益人、申請人的名稱可查對是否本公司承做的案件，若將來可能交由第三者承做時，信用狀宜為可轉讓者。

2. 信用狀的金額應與合約所訂金額相等，過與不足均須澄清或修改以免日後發生困擾和糾紛。

3. 信用狀所列貨品名稱，只需能認可與合約所訂貨品相同即可。

4. 由裝運指示得知有無充裕的時間裝運和裝運有關的條件，同時注意特別條款有無其他裝運要求，若屬不合理者應洽商解決。

5. 由單據方面可知須否於裝運之前應辦檢驗、公證品質證明、投保等有關事項。

步驟三：核對付款有關條件

1. 匯票的被出票人是否為銀行，並注意匯票的付款期限是否與原約定期限一致。

2. 匯票項下的跟單若有無法齊備者，應要求刪除或修改以其他可取得的單據代之。蓋信用狀的押匯業務中有關各方所處理者係單據而非貨品，而且銀行憑表面所示與信用狀條款相符的單據承做付款。

3. 查對有效日期與裝運日期之間的時日合不合理，以免造成逾期失效或趕辦押匯之苦。

4. 查對特別條款有無關於付款事宜。

5. 其他應過目的條款有：銀行承諾條款及適用的信用狀統一慣例的修訂年代，和國際商會頒佈的版本編號，以為日後法律上的依據。

(二)修改信用狀

經過以上的步驟而認可的信用狀，出口廠商即在轉知銀行的通知書回單上簽署以示接受；若拒絕接受時應連同信用狀正本寄轉知銀行。至於信用狀的內容有部分條款不便同意或接受時，宜函知進口商促請修改以利進行準備交貨。一般而言，進出口商雙方均可能有修改信用狀的原因發生。進口商方面於開發信用狀後發覺有誤，例如金額、品名、數量規格等的筆誤，其他內容的誤繕等，在出口商未接獲信用狀之前，得透過原開狀銀行修改，但在出口商接受信用狀之後須經過雙方同意再修改，當然誠實的出口商不會接受與合約略有出入的信用狀，誤繕係非惡意者出口商當然不便拒絕修改的要求。而出口商於接受信用狀之後提出修改的要求者，以裝運指示方面例如分批裝運、裝運日期及有效日期等事項為多。

我國的進口商修改信用狀時有一基本原則，即修改的內容涉及輸入許可證時，宜先修改輸入許可證憑以修改原信用狀的內容，其餘在不逾越外貿有關規定者，只要雙方同意即可逕向原開狀銀行申請修改信用狀。

大直的實踐貿易公司最近一批女用 V 領套頭上衣的訂單，由於未及裝運擬函請進口商展延裝運和有效日期〔例 8－6〕。

〔例 8－6〕

Gentlemen:

Re: Extension for shipment covering
your Order No. M／75

Thank you for your L／C No. 67862 for US$12, 400.00 issued by J. Henry Schroder Bank and Trust Co., New York Under captioned order.

However, after our careful inspection prior to shipment, we found assortment of Ladies V－neck Pullover, was partly mispacked. Therefore, 2 more weeks will be necessary for repacking the items.

Please extend the shipment and expiration dates to January 10, 1998 and January 30, 1998 respectively.

Your approval to the amendment of L／C will be much appreciated.

Yours sincerely,

第八章　交易的履行

第一段：謝謝接獲信用狀，言外之意正準備交貨。

1. L/C No. …for… : ……號信用狀金額……。

2. issued by… : 指開狀銀行。

3. under captioned order : 標題所述訂單。

文意：頃獲貴公司 M/75 號訂單項下由傑亨修銀行所開美金一萬二千四百元的第 67862 號信用狀，謹致謝意。

第二段：裝運之前驗貨發覺配色包裝有誤，須重裝。

1. inspection prior to shipment : 裝運前的檢驗。

2. partly mispacked : 有部分誤裝；部分裝錯。

文意：惟經本公司於裝運前仔細的驗貨結果，顯示女用套頭上衣的配色有部分誤裝，因此重行包裝須時約兩週。

第三段：請展延裝運日期和有效日期。

1. extend…to… : 將……展延至……。

2. respectively : 分別地。

文意：煩請分別展延裝運日期和有效日期至一九九八年元月十日和一九九八年元月三十日。

第四段：敬請惠予修改信用狀作為結尾。

1. approval to the amendment of L/C : 同意修改信用狀。

文意：敬請惠予修改信用狀為禱。

練習問題 •————————————————————————————

一、試述信用狀的開狀方式與其體裁的關係。

二、何謂 SWIFT 方式的信用狀。

三、試參照〔例 8－1〕，歸納〔例 8－2〕的信用狀內容。

四、試參照〔例 8－1〕，將〔例 8－3〕的信用狀內容歸納為十大項。

五、何種項目須於開狀之前申請輸入許可證？

六、出口商接獲國外開來的信用狀，應如何核對其內容？

七、東和貿易公司與日本中外製藥公司之間成交一筆生意，取得中外製藥
　　的銷貨確認函如下〔例 8－7〕。根據東和貿易公司的資料，試代為
　　填製開發信用狀申請書並試算應繳結匯價款和開狀有關手續費。
　　東和貿易公司的資料：

　　(1)申請開狀日期為 Dec. 20, 1998。

　　(2)以航郵開發。

　　(3)商業發票四份。

　　(4)裝箱單四份。

　　(5)全套清潔提單，以中央信託局為受貨人，並通知進口商。

　　(6)進口港為基隆。

　　(7)裝船日期不得遲於 March 5, 1998。

　　(8)不可分批裝運。

　　(9)不得轉運。

　　(10)押匯須於裝船日後五天內為之。

　　(11)信用狀有效期限為 March 10, 1998。

　　(12)國外發生之銀行費用由出口商負擔。

　　(13)通知銀行：Dai－Ichi Kangyo Bank, Tokyo。開發信用狀申請書如
　　　〔例 8－8〕及〔例 8－9〕。

　　(14)開狀時繳交一成保證金。

(15)開狀手續費費率以 0.1% 計算，最低收費新台幣 400 元。

〔例 8－7〕

```
CHUGAI PHARMACEUTICAL CO., LTD.
5－1, 2－Chome, Ukima, Kita－Ku
Tokyo, Japan
SALES CONFIRMATION
                                    Tokyo, Dec. 10, 1997
```

To: Tong Ho Trading Co., Ltd.
100, Chungking South Road, Sec. 3
Taipei, Taiwan, R. O. C.

Dear Sirs,
We confirm having sold to you the following goods on terms and conditions set forth below:

Commodity: Guronsan Tablets.
Specification: 50mg, 120's 05843.
Quantity: 10, 000 boxes.
Price: US¢53 per box FOB vessel Japanese port.
Total amount: US$5, 300. 00.
Packing: in export standard cartons.
Shipment: within 30 days after receipt of L/C, which must be opened by
          end of Dec. 1997.
Destination: Keelung, Taiwan.
Payment: By a prime banker's irrevocable L/C payable against sight draft.
Insurance: Buyer's care.
Remark: Please advise L/C thru Dai－Ichi Kangyo Bank, Tokyo.

                                                Yours, truly,

〔例 8-8〕

開發信用狀申請書及約定書
APPLICATION & AGREEMENT FOR LETTER OF CREDIT

中央信託局   台鑒
To: CENTRAL TRUST OF CHINA

(*31) DATE OF ISSUE _____

(*20) L/C NO. _____

We bound ourselves to the terms of the reverse side, request you to issue an irrevocable documentary credit as follows by: □Airmail 航郵 □Brief Cable 簡電 □Full Cable 全電	(*31D) Expiry date in beneficiary's country for negotiation _____ (有效期限) (*59) Beneficiary (受益人名稱及地址)
(*50)Applicant (申請人名稱及地址) Telephone:	(*32B) Amount: 金額(小寫) (大寫)
Advising Bank (通知銀行，如需指定請填上)	(*41)Credit available with(如需限押，請填上押匯銀行名稱)
(*43P) Partial shipments □allowed □not allowed (*43T) Transhipment □allowed □not allowed (*44A) Shipment Loading on board/dispatch/taking in charge from/at · (*44B) for transportation to (*44C) not later than By □ Vessel □ Air Freight	BY □ NEGOTIATION □ ACCEPTANCE □ PAYMENT □ DEFERRED PAYMENT against presentation of the documents detailed herein (*42C) and of beneficiary's draft drawn □ at sight □at ____ days □after sight □after B/L date for full invoice value on yourselves

(*45A) Evidencing Shipment of: (貨品名稱)

Price Terms: □FAS □FOB/FCA □CFR/CPT □CIF/CIP _____
(請填上地點)
(*46A)Documents Required:
  □ Manually signed commercial invoice in quadruplicate
  □ Packing List in quadruplicate
  □ Full set □ 2/3 set of original clean on board ocean bills of lading made out to order of
    □ the issuing bank □ the applicant, marked "Freight □ Prepaid □ Collect", notifying the applicant with detailed address
  □ Original Clean Air Waybill consigned to □ the issuing bank □ the applicant, marked "Freight □ Prepaid □ Collect", notifying the applicant with detailed address
  □ Insurance Policy or Certificate in duplicate, blank endorsed, issued in the currency of the credit for 110% of the invoice value, with claims payable in Taiwan covering
    □ Institute Cargo Clauses □(A)/ □(B)/ □(C)/ □Air
    □ Institute Strikes Clauses (Cargo)
    □ Institute War Clauses (Cargo)
    □ Others
  □ Beneficiary's Certificate stating that they have forwarded □1/3 set of original B/L and □one complete set of non-negotiable documents directly to the applicant by registered airmail/courier after shipment.

(*47A) Special Instructions:
  □ All documents must bear this credit number.
  □ All banking charges outside of Taiwan are for account of beneficiary.

(*48) Documents to be presented for negotiation within _____ days after the date of shipment but within the validity of the credit.
(*49) Confirmation Instructions: □ Without
           □ Confirm, charges for □ buyer's/ □ beneficiary's account

郵 電 費		結匯成數		本案信用狀各項書表均經審核無誤，並辦妥各項手續，擬准照辦。								
手 續 費		結匯匯率										
保 兌 費				核 准	經 副	理	襄 理	科(課)長	副科(課)長	承 辦	會計科會簽	
合 計												

86. 1. 2,000 L

〔例 8-9〕

約 定 書

前頁開發信用狀之申請倘經貴局核准申請人自願遵守下列各條件

一、關於本信用狀下之匯票及其附屬單據等如經貴局或貴代理行認為在表面上尚屬無訛申請人於匯票提示時應即承兌並依期照付。

二、上項匯票單據等縱或在事後證實其為非真實或屬偽造或有其他瑕疵概與貴局及貴代理行無涉其匯票仍應由申請人照付。

三、本信用狀之傳遞錯誤或遲延或其解釋上之錯誤及關於上述單據或單據所載貨物或貨物之品質或數量或價值等之全部或一部減失或遲逸或未經抵達交貨地以及貨物無論因在海面或陸上運輸中或運抵後或因未經保險或保額不足或因承辦商或任何第三者之阻滯或扣留及其他因素各等情以致喪失或損害時均與貴局或貴代理行無涉在以上任何情形之下該匯票仍應由申請人兌付。

四、與上述匯票及與匯票有關之各項應付款項以及申請人對貴局不論其現已發生或日後發生已到期或尚未到期之其他債務在未清償以前貴局得就本信用狀下所購進之貨物單據及貴得價金視同為自己所有並應連同申請人所有其他財產包括存在貴局及分支機構或貴局所管轄範圍內之保險金存款餘額等均任憑貴局移作上述匯票之共同擔保以備清償票款之用。

五、如上述匯票到期而申請人不能照兌時或貴局因保障本身權益認為必要時貴局得不經通知有權決定將上述財產（包括貨物在內）以公開或其他方式自由變賣就其貴得價金扣除費用後抵償貴局借墊之款毋須另行通知申請人。

六、本信用狀如經展期或重開及修改任何條件申請人對於以上各款願絕對遵守不因展期重開或條件之修改而發生異議。

七、本申請書之簽署人如為二人或二人以上時對於本申請書所列各項條款自當共同連帶及個別負其全部責任並負責向貴局辦理一切結匯手續。

八、申請人對貴局根據本申請書所簽發之信用狀內容所載各項規定及要求單據與本申請書所載者有差異時，該項差異如係貴局所為符合法令、統一慣例或為保障申請人及貴局權益所作之修改者，申請人自當同意，申請人如不同意應於收到信用狀副本聯後一日內通知貴局更正。逾期視為接受貴局簽發之信用狀所發一切內容。

九、在機動匯率之情形下，申請人因文件欠缺或申請書所載內容條款不夠明確或因本信用狀涉及融資手續尚未辦妥或因開狀有關單證之送達已超過貴局之營業時間等不可歸責於貴局之原因致貴局無法及時開狀並結售外匯時，其一切後果及匯率風險概由申請人員責，貴局得以申請人並同意照貴局通知之實際承售匯率多退少補結匯款。又倘貴局因申請人經辦人之口頭或書面要求，已照某日之匯率結購本信用狀下所需外匯，其後因故本信用狀不能開出或延緩開出或開出後信用狀金額未用完致有結餘外匯時，申請人願除對結購匯率不得提出異議外並負責通知貴局辦理退匯，並同意照貴局辦妥退匯日之即期匯率結售貴局。

十、本申請書確與有關當局所發給之輸入許可證內所載各項條件及細則或有關交易文件絕對相符，並已逐一遵守，倘因申請人對於以上任何各點之疏忽致信用狀未能如期開，貴局概不負責。又貴局有刪改本申請書內之任何部份，俾與輸入許可證所載者相符之權，此外申請人應遵守開狀當日有效的國際商會公告之信用狀統一慣例。

本案係　　　　　　股份有限公司購料貨款案

借款契約號碼：

契約有效日期：

借款契約額度：

可動用金額：

本次聲請動用金額：

本信用狀　　　　%金額計　　　　元

由本科上述貨款案項下墊付，其中　　　%

金額　　　　元由該公司自籌結匯。

單證請交授信　　科（課）

申請人

地址：

電話：

核	經副裏理	授信科（課）	科（課）長	副科（課）長	經辦人
定					

會
會計科(課)

貿易英文

158·

Trade 第九章 English

交貨與付款

一、出口簽證、檢驗與報關

出口廠商於接獲信用狀之後，即着手準備貨品的包裝託運之外，其貨品屬於需簽證項目者，應申請輸出許可證(Export Permit)，須檢驗之貨品應先行報請檢驗，合格後始可報關裝運。

1.出口簽證

輸出許可證申請書一套三聯。第一聯為輸出許可證申請書〔例9－1〕，第二聯為輸出許可證(副本)，第三聯為輸出許可證(正本)〔例9－2〕。填寫申請書的要領參考輸出許可證各欄填寫說明(附錄四)。

申請人繕妥申請書後，向國貿局或委託之指定銀行(俗稱簽證銀行)申請，經核准後領回第三聯並應於簽證後三十天內報關。

2.出口檢驗

屬於我國商品檢驗法項下的貨品，須填妥商品輸出報驗申請書〔例9－3〕向商品檢驗局申請檢驗取得合格證書後，始得辦理出口報關。非屬於商品檢驗法項下的貨品，若國外的進口商要求出口商出具政府機構的檢驗證明書時，亦得委請商品檢驗局檢驗。除此之外，雙方得委託指定的公證行檢驗並出具檢驗證明書(Inspection Certificate)〔例9－4〕。

3.產地證明書(Certificate of Origin)

於出口檢驗的同時，出口廠商應進口商的要求，配合進口地報關上的需要，須向經濟部申請核發產地證明書，以證明出口貨品係台灣地區輸出或確實是台灣地區所生產、加工或製造者，此項經濟部核發的產地證明書有兩種：一為優惠關稅產地證明書(Generalized System of Preferences Certificate of Origin Form A)〔例9－12〕；二為一般產地證明書(Certificate of Origin for Taiwan Products)〔例9－13〕。除此之外，商會亦可出具產

〔例 9-1〕

REPUBLIC OF FOREIGN TRADE

第一聯：簽證機構存查聯

輸 出 許 可 證 申 請 書
APPLICATION FOR EXPORT PERMIT

共	頁 第	頁

①申請人：Applicant	③目的地國別 Country of Destination	④轉口港 Transhipment Port
②申請人印鑑：Signature of Applicant （請蓋國際貿易局登記之印鑑）	⑤買主 Buyer ⑥收貨人 Consignee ⑦檢附文件字號：	

⑧ 項次 Item	⑨貨品名稱、規格等 Description of Commodities, etc.	⑩商品分類號列 及檢查號碼 C.C.C. Code	⑪數量及單位 Q'ty & Unit	⑫條件及金額 Terms & Value

簽證機構加註有關規定 Special Conditions	輸出許可證號碼 Export Permit No: 簽證機構簽章及日期 Approving Agency Signature and date

一、本輸出許可證自簽證日起三十天內有效，但簽證機構另有規定者，從其規定。 二、本輸出許可證僅限一次報運，一經報運即為失效，商品分類號列及有簽證機構校 　　正者為準。 三、本申請書所列各項貨品不得分割，倘須分割時須申請換證。	收件號碼 收件日期

貿易英文

162

〔例 9-2〕

輸 出 許 可 證（正本）
EXPORT PERMIT（ORIGINAL）

第三聯：出口人報關用聯

| 共 | 頁 | 第 | 頁 |

①出口人：Exporter	③目的地國別 Country of Destination	④轉口港 Transhipment Port
②出口人印鑑：Signature of Exporter	⑤買 主 Buyer	
	⑥收 貨 人 Consignee	
	⑦檢附文件字號：	
（請蓋國際貿易局登記之印鑑）		

⑧ 項次 Item	⑨貨品名稱、規格等 Description of Commodities, etc.	⑩商品分類號列 及檢查號碼 C.C.C. Code	⑪數量及單位 Q'ty & Unit	⑫條件及金額 Terms & Value

簽證機構加註有關規定 Special Conditions	輸出許可證號碼 Export Permit No:
	簽證機構簽章及日期 Approving Agency Signature and date
海關報單號碼： 海關簽放簽章日期：	

一、本輸出許可證自發證日起三十天內有效，但簽證機構另有規定者，從其規定。
二、本輸出許可證應一次套打，一經塗改即屬失效，商品分類號列蓋有簽證機構校
　　對章者除外。
三、本輸出許可證記有貿易資料，關係商業機密，請予保密，不得外漏或買賣。

| 收件號碼 |
| 收件日期 |

正　　本	商品輸出報驗申請書 APPLICATION		一乙甲俟 枚字字甲

敬啓者：本 公司 擬出口下列產品請予檢驗發證為感
　　　　　　行

Dear Sirs:

　　　　We wish to export the following products,
please make inspection according to your regulations
and issue proper certificate(s).

受理時間 ＿＿＿＿＿＿＿＿

分類字號 ＿＿＿＿＿＿＿＿

申請號碼 ＿＿＿＿＿＿＿＿

1. 申請人（商號）
　 Applicant ＿＿＿＿＿＿＿＿＿＿＿＿＿＿＿＿　蓋章

2. 生　產　者（工廠編號）
　 Producer ＿＿＿＿＿＿＿＿＿＿＿＿＿＿＿＿

3. 輸　出　者
　 Exporter ＿＿＿＿＿＿＿＿＿＿＿＿＿＿＿＿

4. 品　　　　名
　 Commodity ＿＿＿＿＿＿＿＿＿＿＿＿　　中國商品標準號列
　　　　　　　　　　　　　　　　　　　　　　C. C. C. Code ＿＿＿＿＿

5. 規　　　　格
　 Specification ＿＿＿＿＿＿＿＿＿＿＿　　標　誌
　　　　　　　　　　　　　　　　　　　　　　Mark

6. 數　　　　量
　 Quantity ＿＿＿＿＿＿＿＿＿＿＿＿＿

7. 總　淨　量
　 Total net weight ＿＿＿＿＿＿＿＿＿

8. 國別代號
　 Destination ＿＿＿＿＿＿＿＿＿＿＿

9. 檢驗標誌號碼
　 Inspection Label Nos. ＿＿＿＿＿＿＿

10. 檢　驗　日　期
　 Date of inspection ＿＿＿＿＿＿＿＿

輸出價格 F.O.B. ＿＿＿＿＿＿＿＿　　申請人地址及連絡電話 ＿＿＿＿＿＿

外幣換算率 ＿＿＿＿＿＿＿＿＿＿＿　　申請人特別要求 ＿＿＿＿＿＿＿＿

結匯證件號碼 ＿＿＿＿＿＿＿＿＿＿　　貨品堆積地點 ＿＿＿＿＿＿＿＿

科　長（課長）　　　　　　　經辦人

收 費 欄	收費類別	檢驗費	旅　　費	延長作業費	標　識　費	其　他　費
	金　　額					
	收款單號碼					
	稽核人 蓋章	收費人 蓋章				

補　收 費用欄	實際結匯金額 F.O.B.us$	補收金額 U. S. $
	應收檢驗費 N. T. $	補收檢驗費 N. T. $

取 樣	包裝檢查	商標 數量 外觀 內容	重量檢查	淨重 毛重 總淨重	取樣情形	開件數 開件號碼 取樣數量	日期 氣候	月　日　時 晴　陰 雨　雲

7. 100,000

科　長（課長）　　　　　　取樣員

〔例 9-4〕

RFI CORPORATION

WE HEREBY CERTIFY THAT THE SHIPMENT IS
FULLY IN COMPLIANCE WITH THE CONTRACT
REQUIREMENTS IN SPECIFICATIONS,
QUANTITY, QUALITY, PROPER PACKING AND
MARKING.

100 PINE AIRE DRIVE, BAY SHORE, L. I., N. Y. 11706
AREA CODE 516 231-6400
TWX 510 227-6233
FAX 516 231-6465

INSPECTION CERTIFICATE

CERTIFICATE OF TOTAL CONFORMANCE

LC #   DDH /0077
Contract: C460.
LI# /0

CUSTOMER _____ *Coastal Consultants*

PART NUMBER _____ *RF 119*

REVISION _____ *NP*

PURCHASE ORDER # _____ *C460*

QUANTITY _____ *15 pcs.*

DATE CODE _____ *9523*

SERIAL NUMBERS (IF APPLICABLE) _____ *NP*

THIS IS TO CERTIFY THAT ALL PARTS/MATERIALS ON THE ABOVE PURCHASE ORDER HAVE BEEN
MANUFACTURED AND INSPECTED/TESTED, AND ARE IN ACCORDANCE WITH THE SPECIFICATIONS,
DRAWINGS, AND PURCHASE REQUIREMENTS.

IT IS FURTHER CERTIFIED THAT THE PARTS/MATERIALS WERE MANUFACTURED AND INSPECTED/
TESTED IN ACCORDANCE WITH THE FOLLOWING SPECIFICATIONS, NOTES, ETC.

MANUFACTURED DATE: 8/1/94

MIL-SPEC NUMBER _____ *NP*

DRAWING NOTES _____ *NP*

SPECIAL TEST _____ *NP*

THE INSPECTION/TEST REPORTS ARE ON FILE AND ARE AVAILABLE FOR YOUR REVIEW.

_____ Q.C. Manager   *6-5-95*

SIGNATURE AND TITLE   INSPECTION DATE

地證明書〔例9－14〕提供出口廠商的需要。

4.裝運程序

目前的對外貿易交貨的運輸方式分：(1)海運，即出口地與進口地之間以海洋輪船為運輸工具；(2)航運，即以航空器為運輸工具及(3)郵寄，即經由郵包方式運輸者。又由於貨櫃運輸之興起，出、進口兩地之間可以聯合兩種以上的運輸方式，稱為複合運輸(Multimodal Transportation)。由於兩地之間的運輸可以涉及兩種以上的運輸工具，因此 Shipment 一詞被廣泛地解釋為裝運。但我國外銷仍以海運為主，以下即以海運為例，說明其作業情形。

外銷的貨品屬於雜貨類(General Merchandise)者，通常交由定期船(Liner)，而大宗貨物(Bulk Cargo)則委由不定期船(Tramp)運輸出口。以定期船為例，成交的貿易條件為 C&F 或 CIF 條件者，出口廠商應配合船期表選擇適當的船隻(Vessel)向船公司或船務代理行洽訂艙位。若係 FAS 或 FOB 條件時，應將貨品運送至進口商指定的船隻停靠地點備裝。惟若進口商未指定船隻或授權出口商代辦船運時，出口商宜代為接洽。洽訂艙位時填寫託運單(Booking Note)，領取裝貨單(S／O: Shipping Order)〔例9－5〕，即船公司指示船長接受單上所載貨物予以裝載的憑證，亦為出口商辦理報關必備文件之一。俟辦妥報關手續由海關在裝貨單上加蓋關印後，連同貨物運至船邊裝船。裝艙時船公司與出口商(或由報關行代表)派有理貨人員(Tally Clerk)會同檢查裝貨件數及包裝情形，雙方在理貨單(Tally Sheet)上簽字，並俟運費付訖(FAS FOB，C&I 等條件則於卸貨時由進口商支付)後憑以換領提單。

5.貨櫃運輸(Container Shipment)

貨櫃運輸乃配合近代捷運系統的需要而產生的聯合運輸作業(Combined Transport Operation)。貨物裝櫃的方式分為整裝貨櫃(FCL: Full Container Load)與併裝貨櫃(LCL: Less than Full Container Load)。

Shipper

貨主中文名稱：
貨主拾項清開：(若與SHPR相同則不必專列)　FAX:NO:

統一編號：
Consignee

Notify Party
　　　　　　　　　FAX:NO:

WAN HAI LINES LTD.

(CONTAINER SERVICE)

SHIPPING ORDER

	FAX:NO:
Ocean Vessel	Voy.No.
Place of Receipt	Port of Loading
Port of Discharge	Place of Delivery

Please receive for shipment the undermentioned Goods in apparent good order and condition, unless noted below, and sign the accompanying Dock Receipt for same.

Final destination(for the Merchant reference)　BOOKING NO.

Marks & Numbers	Quantity	Description of Packages & Goods	G./N. Weight	Measurement

品名須詳實註明。如僅繕打 General Merchandise 恕無法受理。

①貨名請務必說明。
②超重、整體線請洽換其轉詢材積。
③單件重量超過 5 TONS 者另收
　特殊 SIZE 貨物請先告知本公司。

櫃型／櫃數：

□ 普通櫃　_____ × 20'　_____ × 40'

□ 冷凍櫃　_____ × 20'　_____ × 40'

□ HQ　　_____ × 40'

□ SOC　_____ × 20'　_____ × 40'

□ 其他特殊櫃　_____

SERVICE REQUIRED
務須註明運送方式

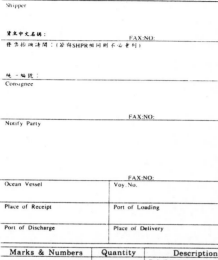

□ 電報放貨

□ 運費證明

□ 船齡證明

□ PREPAID 預付　□ COLLECT 到付

請注意：化學品之類貨物一概不收，
不論具危險性及非危險性。

Particulars Furnished By Shipper of Goods

1. □ FCL/FCL　整櫃貨 整櫃貨
2. □ FCL/LCL　整櫃貨 併櫃貨
3. □ LCL/FCL　併櫃貨 整櫃貨
4. □ LCL/LCL　併櫃貨 併櫃貨

Shippers are requested to note particularly the terms and conditions printed on the back of Dock Receipt.

Please of receiving B(s)/L
領取單證
□ 台北　　□ 台中　　□ 高雄
萬海航運股份有限公司

RECEIVED for shipment

No. of Containers or Packages _____

Receipt: Date _____ Time _____

Place of receipt CFS/CY location _____

請托運人注意收貨單背面各項條款

＊S/O上之內容若有變更請劃出，在結關日前書明或得填妥後交回。
　謝謝。

基中結關：萬海台北總公司　FAX NO. (02)5216000.5632254
台中結關：賴泰代理行　　　FAX NO. (04)6562071.6562069
高雄結關：萬海高雄分公司　FAX NO. (07)2615500.2919006

②

Checker

免費語音查詢電話
080-012321　語音代碼

□□□□□□□ ＊

NAME OF CUSTOMS BROKER

報關行：

電　話：

FAX NO.

S/O NO.

B/L NO.

萬 0008-A-0008-①

前者託運的貨物數量可達整櫃裝滿，而後者則貨物數量不足整櫃。整裝貨櫃採到廠集貨方式，由船公司派拖車將空櫃拖到工廠或裝貨地裝櫃，然後運到碼頭貨櫃場(CY: Container Yard)；至於併裝貨櫃採到站集貨方式，即貨主將貨物運到貨櫃集散站(CFS: Container Freight Station)交由管理員按所有貨主託運的貨依體積、噸位、性質、目的地併裝一櫃。

6.出口報關

出口商得以自己為報關人，亦可委託報關行向海關申報出口。報關前應準備：(1)船公司的裝貨單(S∕O)；(2)輸出許可證正本，但貨物屬於免簽發輸出許可證者，則免附於出口報單；(3)結關日期前準時將貨物運至碼頭倉庫取得進倉證明文件。填寫出口報單〔例9－6〕、〔例9－7〕。收單之後，海關驗貨人員會同報關人員清點件數，開箱抽驗，再經核對查驗結果、稅則號碼、離岸價格、核銷規費證及有關文件之後在裝貨單及報單上加蓋放行專用關防發還報關人，轉送駐碼頭關員或駐倉庫關員核對印鑑後准予將貨物交由船公司裝船。船公司方面則憑裝貨單辦理結關裝船手續。

二、裝船通知(Shipping Advice)

出口商於報關裝運之前或同時，通常須將貨品裝載數量、船名、出口港、開航日期，甚至提單號碼等通知進口商，以海運交貨時特稱之為裝船通知。根據國貿條規(Incoterms)，在 FAS 條件下，賣方須將貨物在買方指定的船隻停泊處交到船邊，並毫不遲延通知買方；而在 FOB、C & F 及 CIF 條件下，賣方須在裝船之後，毫不遲延通知買方貨物已交或裝至船上。尤其是 FAS、FOB 及 C & F 等條件下，進口商須憑出口商的裝船通知以便辦理保險事宜。

實踐貿易公司於十月二十日將 US$5,600 的貨品裝船後，應進口商的要求先以電報通知裝船事宜並寄發電報確認函(Cable Confirmation)如下〔例9－8〕：

〔例9-6〕

關01002
出口報單

類別代號及名稱(6)		聯別	共　　頁	收單
			第　1　頁	

報單（收單關別　出口關別　民國年度　船或關代號　裝貨單或收序號）　收單編號或託運單號碼(13)
號碼
(7)　／　　　／　　　／　　　／　　　／

報關人名稱、簽章	專責人員姓名、簽章	統一編號(8)	海關監管編號(9)	繳	理單編號

貨物　輸出
出售（中、英文）名稱、地址

報關日期（民國）(14)　輸出口岸(15)
　　年　月　日

			離岸價格(16) FOB Value	金額	
				TWD	
				幣別	

案號

(1)		(2)	買方統一編號(11)（及海關監管編號）名稱、地址	運費(17)

檢附文件字號(3)

保險費(18)

貨物存放處所(4)　　運輸方式(5)

加(19)　費用(20)
要

申請沖退原料稅(21)	買方國家及代碼(22)	目的地國家及代碼(23)	出口船（機）名及呼號（班次）(24)		外幣匯率

項次(27)	貨物名稱、出口、、商品分類號列(28)	輸出許可證號碼－項次(29) 輸出規定（主管機關指定代號）(30)	統計號列(31)（檢查號碼）	淨重（公斤）(32) 數量（單位）(33) （統計用）(34)	統計單位(35)	離岸價格(36) FOB Value（新台幣）	統計方式(39)
						(　　　　　)	
						(　　　　　)	
						(　　　　　)	
						(　　　　　)	

總件數(25)	單位	總毛重(公斤)(26)	海關簽註事項		

					商港建設費

標記及貨櫃號碼

推廣貿易服務費

		建檔	補檔		

合　　計

		分估計費	放行		

繳納紀錄

		核發准單	電腦審核		

| | | | | 證明文件核發 | 關別　份數　核　發　紀　錄 |

其他申報事項

		通關方式	(申請)查驗方式		

第九章　交貨與付款

· 169

〔例 9-7〕

關 01002
出口報單

聯別		共＿＿頁
		第＿＿頁

報單（收單關別　出口關別　民國年度　結匯關代號　裝貨單位答予號）號碼 (7)

項次(01)	貨物名稱、品質、規格、製造廠等(02)	商標	輸出許可證號碼一項次(04) 貨品稅則分類號列(03) 稅則號別 統計號列(05) （主管機關指定代號）	檢查號碼(06)	淨重（公斤）(10) 數量（單位）(07) （統計用）(08)	海關機關專用欄	離岸價格(04) FOB Value （新台幣）	統計方式(09)
							( )	
			( )	( )			( )	
							( )	
			( )	( )			( )	
							( )	
			( )	( )			( )	
							( )	
			( )	( )			( )	
							( )	
			( )	( )			( )	
							( )	
			( )	( )			( )	
							( )	
			( )	( )			( )	
							( )	
			( )	( )			( )	
							( )	
			( )	( )			( )	

〔例9－8〕

Gentlemen:

Confirming our telegraph of this date, we have the pleasure in informing you that we have shipped your order No. 201 by M. V. Ever Spring of Evergreen Line sailing on October 10, 1998 from Keelung to Seattle.

Commercial Invoice, Packing List and Bills of Lading each in 2 non－negotiable copies are enclosed for your reference.

To cover this shipment, we have negotiated the documentary draft for US $5, 600. 00 through Central Trust of China, Foreign Dept., Taipei under L/C No. 78910 issued by The International Commercial Bank, Ltd., San Francisco.

We hope to have the opportunity of filling your order again in the near future.

Sincerely yours,

第一段：確認當日所拍發的電報及其內容，即裝船通知內容。

1. Confirm：確認，證實。買賣雙方的交換電報（exchanged cables），於拍發之後，為慎重起見常須追認，而書面的確認或證實書乃 confirmation。

   Confirmation of our cable despatched on October, 5 1998 is enclosed. 隨函檢送十月五日拍發的電文證實書。

   Confirming our telegraph of this date ＝ To comfirm our telegraph of this date 為確認今天拍發的電報。

2. Shipped your order：所訂貨品已裝，所訂貨品已交運。為免繁複，對外貿易常以訂單代替有關貨品的名稱、詳細規格等，若僅交部分貨品，則應提交貨數量，如 250 dozen of your order。

3. Shipped by M. V. Ever Spring：由常春輪裝載，交由常春輪裝載。M. V. ＝motor vessel 內燃機船。冠於船名之前的簡寫，另有 S. S. ＝Steam Ship 輪船。

4. Line：運輸公司。

American President Lines 美國總統輪船公司。

China Air Lines 中華航空公司

文意：本公司茲確認於本日電告貴公司 201 號訂單項下的貨品業已裝載常春輪於十月十日自基隆開往西雅圖。

第二段：寄送發票、裝箱單及提單副本供進口商參考。

1. each in 2 copies：每樣兩份。

2. non－negotiable copies：不可轉讓的副本。

一般提供進口商參考或備用的裝船文件均非為正本，可稱 Copy 或 Copies，但是事關提單因船公司所發提單成套，每套可轉讓(negotiable)的份數或三份或二份(詳第八章)，因此參考用的副本特稱 non－negotiable copy。在此所送發票、裝箱單都是參考用之故，利用 non－negotiable 一詞共用。

文意：隨函檢送商業發票、裝箱單及不可轉讓的提單各兩份，謹供參考。

第三段：通知已辦理出口押匯。

1. to cover this shippment：為求償本次交貨，為求償本次貨款。

2. documentary draft：跟單匯票。

3. negotiate…through…under L／C：將信用狀項下的…向…辦理押匯。

文意：本公司已將舊金山國際商業銀行第七八九一〇號信用狀項下的跟單匯票金額美金五千六百元，向中央信託局外匯業務處辦理押匯以求償本批貨款。

第四段：以爭取新訂單作為結尾。

1. fill your order：供應訂貨，指滿足訂單的要求。

2. near future：最近的將來，不久。

文意：但願不久能再度為貴公司的訂單提供服務為幸。

三、必要單據（Required Documents）

　　出口商將貨物裝運之後，即準備各項有關的單據或信用狀需要的單據辦理押匯，以便取得貨款（Proceeds）。通常需要的單據如下：

㈠發貨時由廠商自行製作的單據

　　1.商業發票（Commercial Invoice）

　　乃交貨的清單，又稱發貨單，爲發貨人（Shipper）向受貨人（Consignee）所開記載裝出貨品的詳細內容的單據，亦爲買賣雙方履約的證據。

　　(1)未成交或未裝運之前，由出口商提供進口商參考者，稱爲預估發票或非正式發票（Proforma Invoice）。

　　(2)裝運後所開掣的發票爲正式的商業發票，又稱裝運發票（Shipping Invoice），簡稱發票（Invoice）。

　　(3)經由輸入國駐在輸出國的領事館簽證者爲簽證發票（Visaed Commercial Invoice）。

發票的內容通常分上下兩欄〔例 9－9 〕。

上欄標明發票（Invoice）字樣並包括信頭、發票編號、日期及交貨情形。

　　　　①信頭：通常將出口商的名稱及地址印妥使用，三角貿易的情形
　　　　　　下，偶有信用狀要求提示無信頭的發票（Neutral Invoice）。

　　　　②標明發票 INVOICE 字樣。

　　　　③發票編號。

　　　　④發票日期。

　　　　⑤裝運情形：包括裝載船名、裝貨港、卸貨港及裝運日期。

　　　　⑥進口商名稱、地址，即貨款的付款人及風險的擔負者。

　　　　⑦出票條款，用以表示貨款係憑信用狀付款。

　　　　⑧嘜頭和件數。

〔例9-9〕

① F H F

TEL: 2221 (10 LINES)　2. T. H ROAD　　TELEX 124 226 TAIPI
　　　　　　　　　　　TAIPEI, TAIWAN. R.O.C.　　CABLE "HEHI TAII

② ③ INVOICE NO: A138
　　　　　I N V O I C E
　　　　　-------------　④ DATE: OCT 31,1996

⑤ 　　　　　　　　　　　　　　　　　SAILING ON OCT 31,1996
SHIPPED PER BUXHOON NO12
FROM TAICHUNG, TAIWAN　　　　TO OSAKA

FOR ACCOUNT AND RISK OF MESSRS. MAS　CO., LTD 4, KAMI
⑥ 　　　　　　　　　　　　　YANAGI-CHO, HIRANO KITA-KU,
　　　　　　　　　　　　　　KYOTO, JAPAN

⑦
DRAWN UNDER L/C NO. 31-0274.　　　　DATED 961007
ISSUED BY THE BANK OF KYOTO, LTD.
　　　　FOREIGN DEPT.

===
MARKS　&　NOS DESCRIPTION OF GOODS　QUANTITY　UNIT PRICE　　　AMOUNT
----
⑧ S　　　　　⑨　　　　　　　　　　CIF OSAKA
(IN DIAMOND)　　"TA HWA " BRAND　　　　⑫
OSAKA
E/R 30/2　　　BLENDED YARN, RAW WHITE
LOT: 630.　　　ON CONE
C/NO. 1-192
MADE IN TAIWAN　AS PER CONTRACT NO.
R.O.C.　　　　　520
　　　　　⑩ 　　　　　　　　⑪ PER LBS　⑫
　　　　19,200.00LBS　　　USD1.5　　USD28,800.0
　　　　---------------　　　　　　　---------------
　　　　19,200.00LBS　　　　　　　　USD28,800.0

⑬ TOTAL ONE HUNDRED NINETY TWO (192) CARTONS ONLY

⑭ SAY:TOTAL CIF VALUE US DOLLARS TWENTY EIGHT
　　　THOUSAND EIGHT HUNDRED ONLY

⑯ 　　　　　　　　　　　　⑮ F. H. F
THIS DOCUMENTS HAS BEEN SIGNED BY
ELECTRONIC METHOD AND AUTHENTICATED.　　W. Z. U

　　　　　　　　　　　　　　　MANAGER

⑨貨品明細。

⑩交貨數量。

⑪單位價格。

⑫交貨總金額包括其貿易條件。

⑬裝運件數。

⑭大寫交貨總金額。

⑮發票人(出口商)簽章。

⑯電子方式簽署之確認。

製作發票,其內容不但須與契約、信用狀及有關單據互相一致,而且必須正確無誤,因此不宜加印 E& OE(Errors and Omission Excepted 有錯必改)字樣。

2.裝箱單(Packing List)

裝箱單又稱包裝單,乃按貨品的花色及重量的不同而逐件詳載的單據,亦稱內容明細表(Specification of Contents),可供出口商報關、驗貨、檢驗及進口商提貨點件之用。裝箱單的內容務必與實際包裝內容符合,而且不可與商業發票、提單等其他單據所載內容有所矛盾。

裝箱單的內容亦分上下欄〔例 9−10〕。

上欄和發票的內容相同,記載交貨情形。

①信頭,即出口商名稱地址。

②標明裝箱單 PACKING LIST 字樣。

③相關發票號碼,裝箱單係記載花色、重量之不同,其內容自應與交貨之清單相符一致,故而常以其發票號碼為對照之用,本身並不編號。

④製單日期。

⑤嘜頭和箱號。

⑥受貨人,本案以進口商為受貨人。

⑦裝運情形,包括裝運日期、裝載船名、裝貨及卸貨港。

〔例 9－10〕

① F   .H.   F

TEL: .2221 (10 LINES) 2 T. H. ROAD   TELEX 124 226 TAIPEI
                       TAIPEI. TAIWAN. R.O.C.   CABLE:" HEHI " TAIPEI

⑤ MARKS   ② PACKING LIST   ③ INVOICE NO: A138
-----------------   -----------   ④ DATE   : OCT 31,1996
                       ⑥
S   CONSIGNEE   : HAS CO., LTD. KAHI
(IN DIAMOND)   . . YANAGI-CHO, HIRANO KITA-KU,
OSAKA   KYOTO, JAPAN
E/R 30/2
LOT: 630   ⑦ SHIPPING DATE : OCT 31,1996
C/NO. 1-192   NAME OF VESSEL : BUXHOON HO12
MADE IN TAIWAN   PORT OF SHIPMENT: TAICHUNG, TAIWAN
R.O.C.   DESTINATION   : OSAKA
===
    NOS   ⑧ DESCRIPTION   ⑨ NET WEIGHT ⑩ GROSS WEIGHT ⑪ MEASUREMENT

                              (LBS)   (KGS)   (KGS)   (M3)

    "TA HWA. ' BRAND

    BLENDED YARN, RAW WHITE
    ON CONE

    AS PER CONTRACT NO.
      520
                                          @45.36   @49.36
   -192   19,200.00   8,709.12   9,477.12

                              19,200.00   8,709.12   9,477.12

⑫ TOTAL   19,200.00 LBS
⑬ TOTAL ONE HUNDRED NINETY TWO (192) CARTONS ONLY

⑮   ⑭ F   H   F
THIS DOCUMENTS HAS BEEN SIGNED BY
ELECTRONIC METHOD AND AUTHENTICATED.
                                    W. Z. U

                                       MANAGER

貿
易
英
文

176·

下欄則記載包裝的內容。

⑧箱內所裝貨品的名稱與規格。

⑨所裝貨品的淨重。

⑩裝箱後的毛重。

⑪容積。以體積計費的貨品應記載容積。

⑫交貨總數。

⑬裝箱總件數。

⑭簽署,必須與相關發票的簽署相符。

⑮電子方式簽署之確認。

3.海關發票 (Customs Invoice)

外銷美國、加拿大、澳洲、紐西蘭及南非共和國等國家的出口廠商,除須提供商業發票之外,尚須提供各該國海關的特定格式的發票,稱爲海關發票,供進口商報關提貨之用。輸往美國則使用特別海關發票 (Special Customs Invoice 簡稱 SCI 或 Form 5515) 如〔例 9 —11〕。

(二)向有關單位申請的單據

1.領事發票 (Consular Invoice)

領事發票又稱領事簽證發票,乃駐在輸出國領事所簽證的特定格式的發票,其作用爲:(1)進口地課徵進口稅的根據,以此杜絕出口商虛報發票上的貨價較實際貨價爲低,而圖利進口商;(2)供作進口貨品原產地之統計資料以防傾銷;(3)領事館得以收取簽證費而增加收入。目前先進國家已先後廢止此項領事發票制度,僅有少數國家仍維持此項制度。

2.產地證明書 (Certificate of Origin)

產地證明書又稱原產地證明書,係證明出口貨品確係屬於該出口國家生產、製造或加工的證明文件。在進口國與特定國家之間訂有互惠關稅協

〔例 9-11〕

DEPARTMENT OF THE TREASURY	SPECIAL CUSTOMS INVOICE	Form Approved.

DEPARTMENT OF THE TREASURY
UNITED STATES CUSTOMS SERVICE
19 U.S.C. 1481, 1482, 1484

SPECIAL CUSTOMS INVOICE
(Use separate invoice for purchased and non-purchased goods.)

Form Approved.
O.M.B No. 48 RO342

1. SELLER

2 DOCUMENT NR. *

3. INVOICE NR AND DATE *

4 REFERENCES

5. CONSIGNEE

6 BUYER if other than consignee

7 ORIGIN OF GOODS

8. NOTIFY PARTY

9. TERMS OF SALE, PAYMENT, AND DISCOUNT

10. ADDITIONAL TRANSPORTATION INFORMATION *

11 CURRENCY USED

12 EXCH RATE (if fixed or agreed)

13 DATE ORDER ACCEPTED

14. MARKS AND NUMBERS ON SHIPPING PACKAGES	15. NUMBER OF PACKAGES	16. FULL DESCRIPTION OF GOODS	17 QUANTITY	UNIT PRICE		20 INVOICE TOTALS
				18. HOME MARKET	19. INVOICE	

21 ☐ If the production of these goods involved furnishing goods or services to the seller (e.g. assists such as dies, molds, tools, engineering work) and the value is not included in the invoice price, check box (21) and explain below.

22. PACKING COSTS

27. DECLARATION OF SELLER/SHIPPER (OR AGENT)

23 OCEAN OR INTERNATIONAL FREIGHT

I declare:

(A) ☐ If there are any rebates, drawbacks or bounties allowed upon the exportation of goods, I have checked box (A) and itemized separately below

(B) ☐ If the goods were not sold or agreed to be sold, I have checked box (B) and have indicated in column 19 the price I would be willing to receive

24 DOMESTIC FREIGHT CHARGES

25 INSURANCE COSTS

I further declare that there is no other invoice differing from this one unless otherwise described below) and that all statements contained in this invoice and declaration are true and correct

(C) SIGNATURE OF SELLER/SHIPPER (OR AGENT)

26 OTHER COSTS (Specify below)

28 THIS SPACE FOR CONTINUING ANSWERS

貿易英文

178·

THIS FORM OF INVOICE REQUIRED GENERALLY IF RATE OF DUTY BASED UPON OR REGULATED BY VALUE OF GOODS AND PURCHASE PRICE OR VALUE OF SHIPMENT EXCEEDS $500 OTHERWISE USE COMMERCIAL INVOICE.

* Not necessary for U.S. Customs purposes

Customs Form 5515 (8·80)

定而進口商如欲享受優惠稅率時，必須提供該特定國家的產地證明書。我國出口商可向商品檢驗局申請核簽一般優惠關稅產地證明書(Generalized System of Preferences Certificate of Origin 俗稱 GSP 或 Form A)〔例 9 — 12 〕，若非爲享受優惠關稅，但國外進口商要求提供官方的產地證明書時，亦可向商品檢驗局申請核發產地證明書(Certificate of Origin for Taiwan Products)〔例 9 —13 〕。如國外開來的信用狀未指名產地證明書的簽發人時，我國出口商得向商會申請簽發產地證明書〔例 9 —14 〕。

3.檢驗證明書(Inspection Certificate)

檢驗證明書係證明裝運貨品，其品質規格等符合買賣契約所規定的文件。進口商爲避免出口商裝運不符約定品質規格的貨品，常要求出口商提供此項證明文件(參照〔例 9—3 〕、〔例 9—4 〕)。此外進口國海關規定時，進口商當在信用狀上要求出口商提供檢驗證明書。檢驗證明書的簽發人由雙方約定，通常有出口國的政府機構、獨立公證行、鑑定人、製造廠商、同業公會、進口商派駐出口地的代表或代理人及出口商本身等。

4.重量體積證明書(Weight / Measurement Certificate)

憑重量或體積買賣的貨品，常需證明裝運貨品的重量或體積符合契約的規定，由公證人或公證行(Surveyor)出具，專爲重量而做的公證報告稱爲重量證明書(Weight Certificate)，而專爲尺寸大小而做的公證報告，屬於體積證明書(Certificate of Measurement)。

5.衛生證明書(Health Certificate)

衛生證明書乃證明進口貨品的清潔或無疾病，不致妨害公共衛生。動植物、食品、藥品等的進口，必須由出口國家衛生機構出具合格證書備驗始准進口。

〔例9-12〕

經濟部

1. Goods consigned from (Exporter's business name, address, country)	Reference No.
	GENERALIZED SYSTEM OF PREFERENCES
	CERTIFICATE OF ORIGIN
	(Combined declaration and certificate)
2. Goods consigned to (Consignee's name, address, country)	FORM A
	Issued in _____ The Republic of China _____
	(country)
	See Notes overleaf
3. Means of transport and route (as far as known)	4. For official use

5. Item number	6. Marks and numbers of packages	7. Number and kind of packages; description of goods	8. Origin criterion (see Notes overleaf)	9. Gross weight or other quantity	10. Number and date of invoices

This certificate shall be invalidated in case of any erasion strikeover and alteration.

11. Certification	12. Declaration by the exporter
It is hereby certified, on the basis of control carried out, that the declaration by the exporter is correct.	The undersigned hereby declares that the above details and statements are correct: that all the goods were produced in _____ Taiwan District, Republic of China _____ (country)
Certifying authority: Bureau of Commodity Inspection & Quarantine Ministry of Economic Affairs	and that they comply with the origin requirements specified for those goods in the Generalized System of Preferences for goods exported to
	_____ (importing country)
Signature of authorized officer Issuing date	Place and date, signature of authorized signatory

86. 9. 223,000份

〔例 9－13〕

中華民國經濟部
MINISTRY OF ECONOMIC AFFAIRS, REPUBLIC OF CHINA

不得轉讓
Non Transferable 1

臺灣區產品輸往＿＿＿＿＿＿＿＿＿＿＿＿＿＿＿＿＿＿＿＿產地證明書

Certificate of Origin for Taiwan Products to be Shipped to＿＿＿＿＿＿＿＿

日 期
Date＿＿＿＿＿＿＿＿＿＿

號 碼
Certificate No.＿＿＿＿＿＿＿＿

1. 出 口 商 名 稱
   Name of exporter＿＿＿＿＿＿＿＿＿＿＿＿＿＿

   登記號碼
   Registration No.＿＿＿＿＿＿＿＿

   地 址
   Address＿＿＿＿＿＿＿＿＿＿＿＿＿＿＿＿＿＿＿＿＿＿＿＿＿＿＿＿＿

2. 外國進口商或收貨人姓名
   Name of the importer or consignee＿＿＿＿＿＿＿＿＿＿＿＿

   地 址
   Address＿＿＿＿＿＿＿＿＿＿＿＿＿＿＿＿＿＿＿＿＿＿＿＿＿＿

3. 茲證明本證書內所列之產品確係臺灣區出產／加工／製造等特此證明
   This is to certify that the merchandise described herein is grown/processed/manufactured in Taiwan
   Origin Republic of China

包裝標誌（嘜頭） Marks & Number	
數 量 Quantity	
貨 名 Description of Goods	
備 註 Remark	

上述貨物已裝載
The above described merchandise was loaded on board＿＿＿＿＿＿＿＿
（輪船或飛機名稱 Name of the carrier）

自＿＿＿＿＿＿＿ 於約於 on/about＿＿＿＿ 目的 destined for＿＿＿＿＿＿
Leaving
（臺灣港口名 Name of Taiwan port）　（日期 Date）　（目的國港口名 Name of the destination port）

中 繼
Via/through＿＿＿＿＿＿＿＿＿＿＿＿ 持有在臺灣所發聯運提單第
（港口名 Name of port）　on a through bill of lading No.＿＿＿＿＿＿＿

issued in Taiwan by＿＿＿＿＿＿＿＿＿＿＿＿＿＿ 於＿＿＿＿ 所發
（承裝船舶（飛機）公司名稱 Name of the shipping (aviation) Co. in Taiwan）　on
（日期 Date）

4. 上述貨品，必須在下列標明之運輸情形之下，先證明書方為有效。
   This certificate shall be valid only under the condition (s) indicated below:

   □ 以直接裝運運至
   Under direct shipment to＿＿＿＿＿＿＿＿＿＿＿＿＿＿＿＿
   （目的國口岸 Name of the destination port）

   □ 憑上述在臺所發聯運提單運至
   Being shipped on a through bill of lading to＿＿＿＿＿＿＿＿
   （目的國口岸 Name of the destination port）

5. 本證自填發之日起六十日內有效，主管機關並得酌予延長之，並改查作廢。
   This certificate shall be valid for a period of 60 days after the issuing date, unless sooner terminated
   or extended by the issuing authorities. It shall be invalid in case of any unauthorized alternation.

經 濟 部
MINISTRY OF ECONOMIC AFFAIRS
For the Minister and
by authorization

〔例 9-14〕

4F., 158 SUNG CHIANG ROAD
TAIPEI 10430, TAIWAN.
台北市松江路156號4樓

TAIWAN CHAMBER OF COMMERCE

TEL:5365455(代表號)
FAX:(02)5211980

產 地 證 明 書
CERTIFICATE OF ORIGIN

ORIGINAL

日期
Date   AUG 7 1995

1. 出 口 商 名 稱
   Name of Exporter/   C T F A
   Manufacturer

   地   址   4 W A STREET, SECTION 1, TAIPEI, TAIWAN R.O.C.
   Address

2. 外國進口商或提貨人姓名
   Name of the Importer or Consignee   TO ORDER

   地   址
   Address

3. 茲證明本證書內所列之產品確係在台灣區生產／加工／製造特給予證明
   This is to certify that the merchandise described is grown/processed/manufactured
   in Taiwan Origin。

包 裝 標 誌 Marks & Numbers	貨 品 名 稱 Description of Goods	數 量 Quantity	備 註 Remarks
W R 15 PCT 50 KGS	W R 15 PCT TOTAL : 250,000 BAGS	N.W.12,500.00MT G.W.12,530.00MT	B/L NO:B -01

上   述   貨   品   已   裝
The above described merchandise was loaded on board

MV "A O A"   ving TAICHUNG
(輪船或飛機名稱Name of the Carrier)   港口名 Name of Taiwan port)

於／約於   AUG. 7 1995   OME
on/about   (目的地港口名Name of the destination port)
        (日期Date)   destined for

4. 本證若未加蓋本會印信或塗改後如未會校對即作無效
   Any absence of the Chamber's official seal and any
   unauthorized alteration shall make this Certificate invalid.

840818

簽發單位 TAIWAN CHAMBER OF COMMERCE
Issued by:

Authorized Signature

貿易英文

(三)通關後領回的單證

1.海運提單(Marine Bills of Lading)

海運提單(以下簡稱提單)係通關裝船後,憑裝貨單向船公司換取的,乃運送人或其代理人簽發,證明運送貨物的收據或裝載於船上,並約定將該項貨物運往目的地,交給提單持有人的有價證券。提單的作用不外:(1)運送人對託運人的書面收據;(2)運送契約的憑證,換言之,運送人與託運人之間,雙方的權利義務以提單為憑證;(3)物權的證書,即提單代表貨物的所有權,提單的持有人對提單上所載貨物,得為法律上使用、買賣、償債、抵押處分的行為,經合法背書,構成物權的轉讓;交付提單與交付貨物的所有權具有同一效力。

(1)提單的主要內容

以萬海航運股份有限公司的提單〔例9-15〕為例,主要的項目如下:

上欄

　①運送人之名稱(Carrier's Name)及提單之標示(Indication of B/L)。

　②提單號碼(B/L No.)。

　③運送契約條款(Terms of B/L)。

　④託運人(Shipper),通常為出口商。

　⑤受貨人(Consignee),參閱(2)提單之種類,本案為記名式提單。

　⑥到埠通知單位(Notify Party),通常為進口商。

　⑦收貨地(Place of Receipt),貨櫃運輸的接管地(出口地)。

　⑧裝載船名(Vessel's Name)。

　⑨裝貨港(Port of Loading)。

　⑩卸貨港(Port of Discharge)。

　⑪交貨地(Place of Delivery),貨櫃運輸的接管地(進口地)。

〔例 9-15〕

④ Shipper

     F   .H.   F

⑤ Consignee

   TO MAS   CO., LTD.

⑥ Notify party carrier not to be responsible for failure to notify

   MAS. CO., LTD. 4, KAMI
   YANAGI-CHO, HIRANO KITA-KU,
   KYOTO, JAPAN

Pre-carriage by		⑦ Place of receipt
		TAICHUNG
⑧ Ocean vessel	Voy No.	⑨ Port of loading
BUXMOON	NO12	KEELUNG TAIWAN
⑩ Port of discharge		⑪ Place of delivery
OSAKA JAPAN		OSAKA

② B/L No.	TAOSC931 B(
S/O NO. 0931	

萬 海 航 運 股 份 有 限 公 司
① **WAN HAI** LINES LTD.
BILL OF LADING

③ RECEIVED by the Carrier from the Shipper in apparent good order and condition unless otherwise indicated herein, the Goods, or the container(s) or package(s) said to contain the cargo herein mentioned, to be carried subject to all the terms and conditions provided for on the face and back of this Bill of Lading by the vessel named herein or any substitute at the Carrier's option and/or other means of transport, from the place of receipt or the port of loading to the port of discharge or the place of delivery shown herein and there to be delivered unto order or assigns.

If required by the Carrier, this Bill of Lading duly endorsed must be surrendered in exchange for the Goods or delivery order.

In accepting this Bill of Lading, the Merchant (as defined by Article 1 on the back hereof) agrees to be bound by all the stipulations, exceptions, terms and conditions on the face and back hereof, whether written, typed, stamped or printed, as fully as if signed by the Merchant, any local custom or privilege to the contrary notwithstanding, and agrees that all agreements or freight engagements for and in connection with the carriage of the Goods are superseded by this Bill of Lading.

In witness whereof, the undersigned, on behalf of Wan Hai Lines, Ltd. the Master and the owner of the Vessel, has signed the number of Bill(s) of Lading stated above, all of this tenor and date, one of which being accomplished, the others to stand void.

(Terms of Bill of Lading continued on the back hereof)

Final destination (for the Merchant reference)

⑫ Container No. Seal No. Marks and Numbers	Number of containers or packages	Kind of packages; Description of goods	⑯ Gross weight Kgs./lbs.	Measurement M³/cft.
FCL TO FCL		⑮ "SHIPPER'S PACK LOAD COUNT & SEAL" "SAID TO CONTAIN"		
WHLU95352 /20'	⑭		KGM 9477.12	
(	1 VAN 192CTNS)"TA.HWA BRAND			
⑬ 8 (IN DIAMOND) OSAKA E/R 30/2 LOT: 639 C/NO.1-192 MADE IN TAIWAN R.O.C.		⑰ BLENDED YARN, RAW WHITE ON CONE (INV.NO. A138) SAY: ONE CONTAINER ONLY ⑱ "FREIGHT PREPAID"		
Total No. of container or packages (in words)			⑲ **ORIGINAL**	

Freight FREIGHT AS ARRANGED	Weight Measurement	Rate	Per	Prepaid	Collect

CHARGES

Carrier's Reference								TOTAL			
Service		Type of Goods									
RC'V	DEL'Y		Freight prepaid at		Freight payable at			Place and date of issue			
1 CY	1 CY	1 ORD	TAIWAN					TAIWAN OCT. 31 1996			
2 CFS	2 CFS	2 REEF	Ex. Rate in NTS		No. of original B(s)/L			⑳ **WAN HAI LINES LTD.**			
3 DOOR	3 DOOR	3 DANG	Yen 27.54	㉒		THREE (3)					
WH	Date OCT. 31 1996		㉓ Laden on board the vessel		Signature						
B/L	BUXMOON NO12 KEELUNG					B/L		By			
								AS CARRIER			

中欄

⑫貨櫃的運輸方式(Type of Move)及貨櫃號碼。本案為整裝/整拆(FCL/FCL)方式。

⑬嘜頭及件號(Marks & Nos.)。

⑭貨櫃件數(No. of Container)。

⑮貨櫃內容之裝載註記。

⑯毛重(Gross Weight)。

⑰貨物名稱(Description of Goods)。

⑱標明運費已付或待收(Indication of Freight Prepaid or To Collect)。

⑲正本之標示(Indication of Original)。

下欄

⑳簽發日期及地點(Date with Place of Issuance)。

㉑簽署(Signature)。本案標明由運送人(As Carrier)簽發。

㉒正本提單份數(No. of Original B/L)。

㉓裝載註記欄(On Board Notation),參考(2)提單的種類。

(2)提單的種類

①按裝運與否而分

已裝提單:即(On Board Bill of Lading 或 Shipped B/L):乃貨物裝載後,才簽發之提單。此種提單的運送契約之第一句,多為如下的文句:

" SHIPPED "in apparent good order by(輪船公司名稱)on board the ship(船名)in or off the port of……(港名)。買方通常都要求賣方提出此種提單。

備裝提單(Received—for—Shipment Bill of Lading 或 Received Bill of Lading):此種提單僅證明貨物已由輪船公司收到備運,其運送契約之開端第一字為" RECEIVED "。依 1993 年修訂的信用狀統一慣例第 23 條之規定,信用狀規定的海運提單,必須載明貨物業已裝載於標示之船舶,故備裝提單尚須船公司在裝載註記欄加簽裝載日期(On Board Date)。

〔例9–15〕爲已加註裝載日期的備裝提單。

　　②按受貨人而分

　　記名式提單（直接提單）（Straight Bill of Lading）：其受貨人一欄，直接記載受貨人之名稱，而不冠以"Order"一字。此種提單在我國、日本及英協各國，均無不得背書轉讓之限制，但在美國及南美諸國，都不能視爲流通證券，以背書方式自由轉讓，因此在國際貿易上不常應用。

　　指示式提單（Order Bill of Lading）：凡提單之受貨人一欄載有"Order"一字者，均爲指示式提單，此種提單記載受貨人之方法通常有：

(a) To order of xxx Bank

(b) To order of Shipper

(c) To order

　　第一種記載方法是依×××銀行之指示決定受貨人，而所謂×××銀行，通常爲開發信用狀的銀行，此際提單須經該銀行之背書，始能辦理提貨手續。

　　第二種記載方法，是依託運人（賣方）之指示決定受貨人。至於第三種記載方法又稱空白抬頭。第二及第三兩種表示方法在實務上均須經託運人之空白背書始爲有效。〔例9–15〕爲直接以信用狀申請人爲受貨人之記名式提單。

　　③按有無批註而分

　　清潔提單（Clean Bill of Lading）：凡提單未載明貨物或包裝有瑕疵之批註者，均爲清潔提單。

　　不潔提單（Unclean Bill of Lading 或 Foul Bill of Lading）：凡提單載有貨物或包裝有瑕疵之批註者，均爲不潔提單。

　　〔例9–15〕爲清潔提單。

　　④其他

　　簡式提單（Short form Bill of Lading）：提單上的全部或部分運送條款係參照該提單以外之來源或單據者。

　　運送承攬業者簽發之提單：即"Freight Forwarders Bill of Lading"。

運送承攬業者係介於運送人(如船公司、航空公司等)與貨主之間，承攬貨物的輸運事務，其本身不自備運輸工具，不實際從事運輸，故對其所簽發之提單，依 1993 年修訂的信用狀統一慣例第 30 條之規定，除信用狀另行授權外，押匯銀行僅接受其表面表明承攬運送人為運送人或複合運送人，並經其簽署或確認之提單。

  2.保險單據(Insurance Documents)

　　保險單則於向保險公司商議保險內容，並填送投保單，經保險公司受理後領回者，如〔例 9 −16〕。投保人應檢視的主要內容有：
　　上欄
　　　①保險公司名稱(Name of Insurance Company)
　　　②保單號碼(Policy No.)
　　　③保單種類(Type of Policy)
　　　④被保險人(Name of Assured)
　　　⑤理賠幣別與地點(Currency & Place of Claims Payable)
　　　⑥相關發票號碼(Invoice No.)
　　　⑦保險金額(Amount Insured)
　　　⑧裝運事項包括：
　　　　裝運船名(Name of Vessel)
　　　　裝運日期(Sailing Date)
　　　　裝卸貨港(Loading & Discharging Ports)
　　中欄
　　　⑨保險的標的(Subject Matter Insured)包括：
　　　　裝載貨品(Cargo)
　　　　總數及件數(Total Quantity & No. of Package)及
　　　　相關信用狀號碼(L / C No.)
　　　⑩保險條款(Coverage)
　　下欄
　　　⑪簽發地點及日期(Issuing Place & Date)

〔例9-16〕

① 第一產物保險股份有限公司
THE FIRST INSURANCE CO.,LTD.

HEAD OFFICE: 54, Chung Hsiao E. Rd., Sec. 1, Taipei, Taiwan, Republic of China
Cable: "INSURANCE" TAIPEI Tel 3913271 (30 Lines), P. O. Box 1835, TAIPEI
Fax: (02) 3921256 3943640 保戶申訴專線電話: (02)3940510
The Schedule

② POLICY NO. 1000H6A555'

③ MARINE CARGO POLICY

⑤ *Claim, if any, payable in* USD *currency* at OSAKA

④ ASSURED   F   H   F

CLAIM AGENT :
CORNES & CO., LTD.
TEL:(078)332-3421,3321155
TLX:5622310 CORKOB J.
OSAKA
JAPAN

⑥ Invoice No.   A138

⑦ Amount Insured   (USD31,680.00) US DOLLARS THIRTY ONE
THOUSAND SIX HUNDRED EIGHTY ONLY.

⑧ Ship or Vessel   BUXHOOM NO12
Sailing on or about OCT. 24, 1996
From   TAICHUNG TAIWAN TO OSAKA

Any claim documents should be presented
through our appointed agent.

⑨ SUBJECT-MATTER INSURED
"TA HWA   " BRAND

BLENDED YARN, RAW WHITE
ON CONE

TOTAL: 19,200 LBS
192 CARTONS IN CONTAINER
L/C NO. 31-0274

⑩ Conditions
Subject to the following clauses as per back hereof
Covering Marine Risks
Institute Radioactive Contamination Exclusion Clause
Institute Replacement Clause (applying to machinery)

INSTITUTE CARGO CLAUSES (ALL RISKS)
INSTITUTE WAR CLAUSES (CARGO)
INSTITUTE STRIKES RIOTS AND CIVIL COMMOTIONS CLAUSES
SUBJECT TO THEFT, PILFERAGE AND NON-DELIVERY
CLAUSES AS PER BACK HEREOF.
IRRESPECTIVE OF PERCENTAGE

Marks and Numbers as per Invoice No. specified above.   Valued at the same as Amount insured.

⑪ Place and Date Signed in TAIPEI   OCT. 23, 1996   ⑫ IN DUPLICATE

ORIGINAL

Not valid unless countersigned by...................   For THE FIRST INSURANCE CO., LTD.
A   MARINE UNDERWRITING DEPT.
W-01-101   ⑬   Chen Ann Lee
                                                        Chairman ⑭

⑫發行份數

⑬副署 (Countersigned)

⑭簽署 (Signature)

四、出口押匯 (Negotiation)

1.申請出口押匯

押匯原指匯票的轉讓行為，即出售或購買匯票的行為，惟在我國不論匯票的被出票人亦或信用狀有無要求提示匯票，一般係指出口商在信用狀項下向外匯銀行取得出口貨款，此際墊付貨款的銀行稱為押匯銀行 (Negotiating Bank)。押匯銀行憑出口商的信用訂定其承做押匯的額度，並根據開狀銀行的信用及出口商所提示符合信用狀條款的單據決定墊付款項。因此第一次申請出口押匯須先經過銀行的徵信，填具質押權利總設定書 (General Letter of Hypothecation)〔例9－17〕並留存印鑑卡。出口商與銀行建立了出口押匯業務的往來關係之後即可逐筆開製匯票，填寫申請書〔例9－18〕，附送有關單據辦理出口押匯。

2.開製匯票

出口押匯所需匯票通常由押匯銀行提供，出口商照信用狀的匯票條款繕妥即可，茲舉實例如下：

台北市的 FHF 公司於接獲可分批裝運的信用狀號碼31－0274〔例8－3〕後於十月三十一日裝運價值美金 28, 800. 00 元的貨。該信用狀上有關匯票條款如〔例8－3，42C，42A〕。則本案的匯票〔例9－19〕可依序繕打如下：

(1)匯票金額：照發票金額 US$28, 800. 00 打出小寫金額。

(2)製票日期：通常與發票日期同一日。

(3)本案付款期限為即期 (Sight)，故 At 與 Sight 兩字之間不留白以示

At Sight。

(4)匯票的受款人(Payee)實務上均指定受理押匯之銀行。本案為中央信託局，外匯業務處。

(5)標出匯票的大寫金額。

(6)標明憑信用狀出具之匯票。內容應包括信用狀號碼、開狀日期及開狀銀行名稱。

(7)被出票人：本案依信用狀之規定〔例 8 −3，42A〕打出 BOK-FJPJZ。

(8)出票人：為受益人 FHF 並簽署。

匯票繕妥後，將信用狀所要求的有關單據收齊附在匯票之下。

3.附送文件

申請出口押匯時隨申請書附送的文件有

(1)信用狀正本，以詳電開發者附該電文即可，但以簡電開發者，須併
　　附電報確認書。

(2)匯票。

(3)信用狀項下的必要單據。

(4)其他

　　①佣金申請書：出口商所需支付國外代理商佣金，得自貨款外匯中
　　　扣付，此際應附佣金申請書。

　　②轉讓書(Letter of Assignment)：押匯申請人係信用狀的受讓人
　　　(Transferee)，而未經銀行辦理轉讓手續時，須提示原受益人的
　　　轉讓書以資證明。

　　③大陸出口台灣押匯的案件，須填報「大陸出口台灣押匯申報表」
　　　〔例 9 −20〕。

4.領取押匯款

出口商所提示的有關單據經押匯銀行審查後，無瑕疵(Without Discrepancies)者，換言之，所有單據均符合信用狀的條款時，押匯銀行即按當日買入匯率折付新台幣，其計算式如下：

A. 押匯金額：按申請書金額亦即匯票金額。

(1)代扣代理商佣金：按佣金申請書的金額。

B. 成交金額＝〔A－(1)〕×〔當日買入匯率〕

(2)減　出口押匯手續費＝A×1‰×〔當日買入匯率〕(最低費用 NT$500；轉押匯時按 2‰計最低 NT$1000)

(3)減　郵費，分港澳地區、亞洲地區、歐美及其他地區，由各銀行按地區酌收。

(4)減　匯費：由各銀行按件酌收。

(5)減　利息〔自押匯銀行墊付貨款至求償貨款進帳之間的利息，目前各銀行均按幣別計收，如港幣、新加坡幣及日幣通常收七天利息，其他幣別則預收十二天〕。

(6)減　其他費用。

C. 押匯淨額＝B－(2)－(3)－(4)－(5)－(6)

若出口商所提示的押匯單據有瑕疵(With Discrepancies)，則押匯銀行基於本身債權的安全，往往不願意無條件讓購跟單匯票，視瑕疵的輕重、客戶的信用與往來關係，以及開狀銀行、進口商的信用狀況等而擇一處理方式，常見的處理方式有：

(1)電報押匯：將瑕疵內容電知開狀銀行，並請示可否押匯或付款，俟開狀銀行回電同意後給付押匯款。

(2)憑出口商出具的損害賠償書(Letter of Indemnity)〔例 9－21〕先行給付，若押匯後進口商拒絕付款時，憑此向出口商追索該筆款項。

(3)託收：將匯票及有關單據寄往開狀銀行，俟收妥開狀銀行之同意接受該項瑕疵後，再給付押匯款。

GENERAL LETTER OF HYPOTHECATION
質押權利總設定書

To: CENERAL TRUST OF CHINA

      TAIWAN.

中央信託局 公鑒：

    ㈠ 茲因 貴局隨時可能墊付或押匯設定質權書人（以下簡稱立約人）所簽發之押匯匯票（國內或國外），或背書之押匯匯票，爰經雙方協議：凡本書中所載各條款，均應認爲永久繼續有效，隨時適用，凡立約人所簽發或背書之押匯匯票，無論其爲直接或經手他人押匯與 貴局，均須一律看待，有如每次墊付，或押匯，均與重新簽訂本書，有同一效力。

    1.-As you may from time to time advance or negotiate for us Bill or Bills of Exchange (Inland or Foreign) drawn or endorsed by us, with collateral securities, it is hereby agreed between us that stipulations contained in this Memorandum shall be deemed to be continuing and ambulatory, and shall apply to all cases in which such Bills of Exchange may at any time, either directly or through other persons, by negotiated with you by us as if this Memorandum were signed by us on each occasion of such advance or negotiation.

    ㈡ 茲願提供附屬於匯票之貨運單據及其有關貨物與 貴局爲擔保物，以擔保 貴局墊付或押匯立約人所簽發或背書之匯票票款，利息及其有關一切費用。

    2.-Documents attached to the Bill or Bills and the title of relative goods shall be transferred to your bank as collateral security for the advance and negotiation of our Bill or Bills, and for the payment of interest, all cost and other expenses incurred in connection therewith.

    ㈢ 茲授權 貴局之任何經理，或代理人，或上述匯票現在持有人，得將（但非必需之行爲）該匯票擔保品物所有水險，並包括搶劫擄掠及岸上火災等險，自行投保，所有保險費及有關費用，均得加入在內，歸立約人負擔。又 貴局對於該項票據，及其擔保貨物連同以上費用，均享有優先受償權，並得不問其他背書人應負之任何責任，逕行處分取償，或爲代付保險費及其他費用之人取償。並得變賣部份擔保物，以付必須之運費、保險費及其他費用。同時 貴局得照普通商家代理人之事例，代立約人辦理一切應辦事件。立約人當依照付款人，或承兌人之指示，將貨移放于公家或私人之碼頭或倉庫，倘 貴局對於該指定之碼頭或倉庫並無反對之表示。

    3.-We authorize you, or any of your Managers, or Agents, or the Holders for the time being of any such Bill or Bills as aforesaid (but not so as to make in imperative) to insure any goods forming the collateral security for any such Bill or Bills of Exchange against sea risk, including loss by capture, and also against loss by fire on shore, and to add the premiums and expense of such insurances to the amount chargeable to us in respect of such Bill or Bills, and to take recourse upon such goods in priority to any other claims thereon, or against us, without prejudice to any claim against any endorser-or endorsers of the said Bills, for the purpose of reimbursing yourselves, or other person or persons paying the same, the amount of such premiums and expenses, and also to sell any portion of such goods which may be necessary for payment of freight, insurance, and expenses and generally to take such measures and make such charges for commission and to be accountable in such manner, but not further or otherwise than as in ordinary cases between a merchant and his correspondent. And we consent to the goods being warehoused at any public or private wharf or warehouse selected by the Drawees or Acceptors of the Bills, unless you offer any objection to such wharf or warehouse.

（四） 兹授權　貴局或　貴局之任何經理或代理人，或上述匯票持有人，均可接受付款人附有條件之承兑，于票據到期日及票款付清後，　貴局得將隨同匯票作爲擔保之附屬單據，交與付款人，或承兑人。此種授權亦可適用于爲付款人信用之第三者承兑，惟付款人於付款或承兑前已停止支付，或宣告破產，或清理時，則應按照以下所載第六款辦理。

4.-We hereby also authorize you, or any of your Managers, or Agents, or the Holders for the time being of any Bill or Bills of Exchange as aforesaid, to take conditional acceptances to all or any of such Bill, to the effect that, on payment thereof at maturity, the Documents handed to you as collateral security for the due payment of any such Bill or Bills shall be delivered to the Drawees or Acceptors thereof, and such authorization shall be taken to extend to cases of acceptance for honour, subject nevertheless to the power next hereinafter given in article six in case the Drawee shall suspend payment, become bankrupt, or go into liquidation during the currency of any such Bill or Bills.

（五） 兹授權　貴局，凡經　貴局，或匯票承兑人或其代表人，認爲適當，在匯票到期以前無論何時　貴局可將貨物分批交付與任何人，（但非必寓之行爲）惟交付貨物之全部，或一部份時，須支付相當金額，且其金額須與發票上所開列之貨價，或與所擔保之票據所載金額成合理之比例。

5.-We further authorize you, (but not so as to make it imperative), at any time or times before the maturity of any Bill or Bills Exchange as aforesaid, to grant a partial delivery or partial deliveries of such goods, in such manners as you or the Acceptors of such Bill or Bills of Exchange or their representatives may think desirable, to any person or persons on payment of a proportionate amount of the invoice cost of such goods, or of the Bill or Bills of Exchange drawn against same.

（六） 兹再授權　貴局或　貴局之任何經理，或代理人，或匯票現在持有人，于匯票提示而被承兑人拒絕承兑或于匯票到期而被付款人拒絕支付，立約人拋棄作成拒絕證書之要求，對於上述拒絕承兑而支付，或在票據到期前，付款人或承兑人停止支付，或宣告破產，或採取清理步驟時，不論匯票是否已經承兑人附條件承兑或絕對承兑，　貴局均得將該匯票擔保品之全部，或一部份，按照　貴局，或票據持有人，認爲適當之方法，將其變賣，並將所得價款，除去通常手續費用，及佣金外，以之支付該票款及其匯費，倘有餘額，得由　貴局，或票據持有人，以之清償立約人之其他票據（不論其有無擔保），或對　貴局之欠款，或對　貴局負有結算責任之其他方面欠款，凡遇保險貨物發生滅失，立約人授權　貴局，得依照保險單取償，並扣除手續費用，與處分變賣其他貨物情形同，將其所餘淨額按照上述所開辦法加以處理。

6.-We further authorize you, or any of your Manager, or Agents or the Holders for the time being of any Bill or Bills of Exchange as aforesaid, on default being made in acceptance on presentation or in payment at maturity, or any of such Bill or Bills, to waive protest and in case of such default or of the Drawees or Acceptors suspending payment, becoming bankrupt, or taking any steps whatever towards entering into liquidation during the currency of any such Bill or Bills, and whether accepted conditionally or absolutely to sell all or any part of the goods forming the collateral security for the payment thereof at such times and in such manner as you or such Holders may deem fit, and, after deduction usual commission and charges, to apply the net proceeds in payment of such Bill or Bills with re-exchange and charges the balance, if any, to be placed at your or their option against any other of our Bills, secured or otherwise which may be in your or their bands, or any other debt or liability of ours to you, or them, and subject thereto, to be accounted for to the proper parties. In case of loss at any time of goods insured we authorize you, or the Holders thereof, to realize the policy or policies and charge the same commission on the proceeds upon a sale of goods, and to apply the net proceeds, after such deductions as aforesaid, in the manner hereinberfore lastly provided.

(七) 如貨物變賣所得價款淨額不足以償付上開匯票所載金額，（包活當時匯兌市價折合之損耗），茲授權　貴局，或　貴局之任何經理，代理人，或票據持有人，對于不足之數，得向立約人發出匯票取償，但不影響該不足之數向其他背書人之追索權。茲總諒解，凡　貴局，或票據持有人，所出之帳單，即為變賣貨物，已經受有損失之憑證，立約人于該項匯票之提示後，當即如數照付。

7.-In case the net proceeds of such goods shall be insufficient to pay the amount of any such Bill or Bills, with re-exchanges and charges, we authorize you, or any of your Managers, or Agents, or the Holders for the time being of such Bill or Bills as the case may be, to draw on us for the deficiency, without prejudice nevertheless to any claim against any endorser or endorsers of the said Bills for recovery of same or any deficiency on the same; and we engage to honour such Drafts on presentation, it being understood that the Account Current rendered by you or by such Holders shall be sufficient proof of sale and loss.

(八) 不論變賣貨物之情事將否發生，茲授權　貴局，或　貴局之任何經理、代理人、或票據持有人，均得於匯票到期之前，接受付款人或承兌人付款之要求，並於付款後將提單及其他貨運單據等，交與付款人或承兌人，倘　貴局或票據持有人准其提前支付時，並得按照票據支付地之通常利率，計算回扣。

8.-We further authorize you, or any of your Managers, or Agents, or the Holders for the time being of any such Bill or Bills as aforesaid, whether the aforesaid Power of Sale shall or shall not have arisen, at any time before the maturity of any such Bill or Bills, to accept, payment from the Drawees or Acceptors thereof, if requested so to do, and on payment to deliver the Bills or Bills are to allow a discount thereon, at the customary rate of rebate in the place where such Bill or Bills are payable.

(九) 倘匯票付款人拒絕承兌或付款，或匯票到期前擔保貨品業已運抵目的港口，立約人授權　貴局或　貴局之通匯行辦理該匯票擔保品之卸貨、報關、存倉、保險等，　貴局或　貴局之通匯行認為維護此等貨品必要之任何措施。辦理上項措施所發生之有關費用以及卸貨、報關、存倉及保險各從業人員過失、戰爭、天災或其他不可抗力因素所引起之任何損害悉歸立約人負擔。

9.-Should the drawee of the Bill or Bills reject acceptance or payment of the said Bill or Bills, or should the collateral Goods arrive before the date of maturity of such Bill or Bills, we authorize your bank or your correspondent to unload, clear, warehouse the Goods, effect insurance thereon and do any and all other acts which your Bank or your correspondent clear, warehouse the Goods effect insurance theorn and do any and all other acts which your Bank or your correspondent may deem necessary for the proper maintenance of the said Goods. In these cases, not only the expenses and cost incurred in the course of the above acts, but also any damage caused by those people or parties who deal with the unloading clearance, warehousing and insurance in good or bad faith or by reason of war, natural disasters or any other Act of God shall be paid by us.

(十) 押匯匯票經　貴局押匯後，倘因匯票附屬單據與信用狀所規定條件不符或其他任何理由而遭　貴局之貼現銀行或通匯行拒絕處理，或受開狀銀行拒付，或貨物或交付或其他場合被發覺貨物之品質、數量有差異等情事時或其他任何理由致遭對方拒收，立約人願意負全責，一經　貴局通知，隨時償付　貴局匯票金額，利息與其他一切附隨費用，

立約人仍授權　貴局，倘　貴局或　貴局之通匯銀行認爲必要時，不必通知立約人，　貴局可向信用狀開狀銀行，或承兌銀行提出保證書，對此項保證，立約人願意負一切責任。

10.-Should the Bill or Bills negotiated by your bank be refused handling or processing by your discounting bank or correspondent,or unpaid by issuing bank owing to some discrepancy in the Bill or Bills or the Documents attached thereto with the terms and conditions of the Letter of Credit or for any other reasons, or should the acceptance of the shipped goods be refused because of divergence of quality, quantity etc. of the said goods, or for any other reasons, discovered by the interested party or parties upon delivery or any other occasions, we shall take full responsibility thereof and reimburse you at any time the amount of such Bill or Bills, interest and other incidental charges incurred. We further authorize your bank to tender a letter of guarantee to the issuing bank or the accepting bank under the Letter of Credit, without any notification to us in case your bank or your correspondent deems it fit to do so, and we solely shall be held liable for the guarantee thus offered.

(十一) 倘因匯票付款人，信用狀開狀銀行，承兌銀行或保兌銀行無力償付債務，受破產宣告、查封、假扣押、假處分、拍賣等情事時，或因自請宣告破產或和解時，一經　貴局通知，立約人願意償付　貴局匯票金額，利息以及附隨之一切費用。

11.-Should the drawees of our Bill or Bills or the issuing, accepting or confirming banks of the relative Letter of Credit become insolvent, or bankrupt, be seized, provisionally seized, provisionally disposed of, or offered for auction, or even, should the drawees or the above banks apply for bankruptcy or settlement by composition, we agree to pay you upon your notice the total amount of our Bill or Bills with interest and other additional charges.

(十二) 倘係承兌後交付貨運單據之匯票，立約人授權　貴局，將附隨該匯票作爲擔保品之貨運單據，于承兌人承兌該匯票後，交與承兌人。在此情形之下，倘因該匯票到期而承兌人不予付款，則凡因此而發生之結果，均由立約人負其責任。立約人當將該匯票所欠之全部款項，或一部份款項，及因此而增加之匯費及手續費，如數償退　貴局。並擔保貴局不因此而受任何損害。

12.-In case of D/A bills we authorize you to deliver the documents to the acceptors against their acceptance of the Bill or Bills drawn on them. In such a case we undertake to hold you harmless from any consequence that may arise by your so doing and to pay you the amount or any balance of bill with re-exchange and charges if the acceptors should make any default in payment at maturity.

(十三) 倘因押匯匯票所應具備之要項欠缺，以致匯票債權不成立，或因時效，或手續欠全而導致匯票債權消滅時，立約人仍願意償付　貴局匯票票面金額，連同匯票期滿前後所孳生之利息，以及附隨之一切費用。

13.-Should the right of claim on the Bill or Bills be not validly instituted on account of any formal defect, or should it become extinct owing to the default of safeguarding procedure or presentation. We agree to reimburse you for the amount equivalent to the face value of the Bill or Bills, interest incurred thereon before after maturity and other incidental charges incurred in this connection.

(十四) 茲雙方同意，倘押匯匯票因外來干預致不獲付款人承兌，不獲付款人或承兌人付款，或因當地法律規章或其他任何理由致使匯票無法付款，押匯款無從匯付　貴局時，不論該項匯票與（或）附屬單據是否退還，一經　貴局通知，立約人願意立即償付匯票金額、利息及附隨之一切費用，　貴局如須增加擔保品，立約人亦願意提供絕無任何異議。

14.-We hereby agree that, should the Bill or Bills be not accepted by the drawees or not paid by the drawees or acceptors by intervention,or should it happen that the Bill or Bills are not paid or the proceeds thereof are not transferred to you because of the local laws or regulations or for any other reasons, we shall pay the amount of the Bill or Bills with interest and other incidental charges incurred as soon as you inform us in this connection by cable or by mail, notwithstanding no return of the Bill or Bills and/or documents. Should you demand any additional security of us at same time, it shall be given by us without any objection.

(圭) 本設定書所稱之押匯或貼現，係指 貴局根據立約人所提出之開狀銀行所開信用狀之條款，審核立約人所提示之跟單匯票或單據，由 貴局墊付上開跟單匯票或單據項下之款項。 貴局於墊付後如未能於參個月內向開狀銀行或付款銀行收回墊款，不論任何原因，立約人均無條件願於接到 貴局通知時，立即償還 貴局墊款本息。

15.-The word of negotiation as described in this Memorandum means that after checking the documentary drafts or documents presented by us, in accordance with the terms and conditions of the letters of credit issued by the issuing bank, you make advanced payments under the above-mentioned documentary drafts or documents to us. If you fail to collect the reimbursements thereunder within three months after the dates of you advanced payments, we uncondiitonally undertake, regardless of the reasons whatsoever, to refund you the principals of your advanced payments with interests immediately upon receipt of your notice to that effect.

(盂) 立約人同意，倘 貴局依照立約人之申請，押匯立約人根據「憑單據付款信用狀」（**Payment against documents credit**,亦即不寫受益人簽發匯票之信用狀）所提示之貨運單據後，此項單據不論任何理由遭受 貴局之通匯行拒絕處理，或受開狀銀行拒付，一經 貴局通知立約人即時償付單據金額，利息及其他一切附隨費用，並擔保 貴局不因押匯本項單據而蒙受任何損害。

16.-We agree that in case we request your bank to negotiate shipping documents made out under letters of credit which are payable against shipping documents without producing drafts, and in the event of these shipping documents being refused processing, handling, or dealing by your correspondents, or refused payment by the issuing banks for some or any reasons, we shall pay on demand the full amount of the documents together with interest and relative expenses as required by you, and shall undertake to hold your free and harmless from any loss or damage which may occur in connection herewith.

(圶) 立約人授權 貴局或 貴局之通匯行，以 貴局或 貴局之通匯行認為適合之任何方法寄送押匯匯票與（或）附屬單據。

17.-We authorize your bank or your correspondent to send the Bill or Bills and/or Documents to the place of payment by any method as you or your correspondent deems fit.

(共) 倘押匯匯票與（或）附屬票據在寄送中毀損或遺失，或視為已毀損或遺失，或因誤送等意外情事，致令遲延寄達付款地時得不必經任何法律手續，一經 貴局通知，立約人願意根據 貴局帳簿之記錄，作成新押匯匯票連同新附屬單據提供與 貴局，或隨 貴局之指示，立即償付 貴局匯票金額，以及附隨之一切費用。

18.-Should the Bill or Bills and/or documents be destroyed or lost in transit, or assumed as such, or their arrival at the place of payment is much delayed by accident such as mistransportation, a new Bill or Bills, and new Documents shall be presented to your bank by us according to your record book, at your demand without any legal procedures, or alternatively, at your option, the amount of the Bill or Bills, with all expenses, shall be paid to you by us.

(圥) 在匯票或其他任何單據上所簽蓋之立約人簽章或所寫文字， 貴局如認為與存驗於 貴局者相符，或與立約人曾經使用之匯票或其他單據相符時，即使其係偽造變造或被盜用，立約人仍願意負責，並償付 貴局因此而蒙受之損害。

19.-We shall be responsible for the signature, seal or writing used on the Bill or Bills or any other documents accepted by you even though the signature, seal or writing is a forged or stolen one, should you have concluded the same to be identical with those submitted to you by us beforehand or those used on previous Bill or Bills or any other Documents. Any damage, caused you therefrom, shall be paid by us upon your notice.

（〒）　本設定書所稱利息係指　貴局押匯日起至償付日止，按　貴局檀定出押息利率計收。

20.-The term of interest as described herein will be deemed to be calculated at the rate of interest of negotiation regulated by your bank, for the period of time from negotiation to payment.

（二）　立約人同意，一經　貴局請求，即應更換、分割或合併任何經立約人簽署並交由　貴局收執作爲擔保物之票據。

21.-We hereby agree, upon the request of the Bank, to renew, divide or combine any negotiable instruments issued and signed by us and which have been deposited with the Bank as the Securities.

（三）　立約人同意　貴局得就立約人所有財產包括存於　貴局及分支機構或　貴局所管轄範圍內之保證金、存款餘額等均任憑　貴局移作共同擔保品以清償任何現已發生或日後發生已經到期尚未到期之任何未清償債務。

22.-We agree that all our property including securities and deposit balances which may now or hereafter be in your or your branches'possession of otherwise subject or your control shall be deemed to be collateral security for the payment of any indebtedness and liability now existing or hereafter created or incurred by us to your due or not due.

（三）　有關匯票單據與（或）擔保貨物之訴訟，以立約人申請押匯之　貴局營業所所在地地方法院爲管轄法院。

23.-The jurisdication of a judicial court regarding any legal action on our Bill or Bills or documents and/or collateral goods shall be excuted at the District at the location of your office where we submitted such Bill or Bills for negotiation or purchase.

（四）　立約人願遵守國際商會所刊佈「信用狀統一慣例」，並視爲其本設定書之一部份。

24.-We will observe the ''Uniform Customs and Practice for Documentary Credits'' fixed by the International Chamber of Commerce, and deem it as a part of this Letter.

（五）　兹更經雙方協議諒解，凡　貴局所有對於票據上，因退票而發生之一切權利，不因將擔保品之交付與　貴局而受任何影響，亦不因對擔保之任何處分，而影響　貴局對立約人所欠款項範圍以內，在擔保品上占有之物權。此外關於立約人店號莊行公司，因股東、合夥人之死亡、退夥，或加入新夥，或隨時而發生之其他人事變動。換言之，不論本處名稱、牌號、及內部組織之如何變更，凡在立約人繼續營業之時，本文所授權限及其設定，當繼續有效。凡每次立約人匯票經　貴局押匯或墊付，均應認爲立約人又將已經訂立之本書重新訂立。兹又經雙方諒解，凡因　貴局所僱用之居間人或拍賣行之違約行爲而發生之結果，　貴局對於立約人並不負任何責任。此據。

25.-Lastly, it is mutually agreed that the delivery of such collateral securities to you shall not prejudice your rights on any of such Bills in case of dishonour, nor shall any recourse taken thereon affect your title to such securities to the extent of our liability to you as above, and that notwithstanding any alteration by death, retirement, introduction of new partncrs or otherwise in the persons from time to time constituting our firm or the style of our firm under which the business at present carried on by us may be from time to time continued, this Letter and the powers and authorities hereby given are to hold goods as the Agreement with you on the part of the firm as aforesaid, and that each negotiation of a Bill or Bills bereunder is to be treated as a renewal by or on behalf of the firm as then existing of the terms of this Agreement. It is also agreed that you are not to be responsible for the default of any Broker or Auctioneer employed by you for any purpose.

中　華　民　國　　　　　　年　　　　　月　　　　　日

Dated .... this .... day of ....

One Thousand Nine Hundred and ....

核對印鑑	公　司　印　鑑	負責人簽名
核對日期：		
核對人：		

....

質押權利總設定書人簽章　　　　　Signature

地　址：
Address:

電　話：
Telephone:

　　　茲連帶保證上開立約人完全履行本書內所載各項條款之義務並拋棄先訴抗辯權及民法上一切抗辯之權利合為保證如上。

　　　We hereby jointly and severally guarantee the fulfilment of the above, expressly waiving our right to take previous action in Court and all civial right to debate.

對保	連帶保證人簽章
核對人：	

....

連帶保證人簽章　　　　　Guarantor

地　址：
Address:

〔例9-18〕

出　口　押　匯　申　請　書　編 號
APPLICATION FOR NEGOTIATION OF DRAFTS UNDER LC.　BP. No. _____

中央信託局外匯業務處
To: CENTRAL TRUST OF CHINA
FOREIGN DEPARTMENT
Taipei

日 期
Date: _____

Dear Sirs:

茲檢送附本公司所開滙票第　　　　　號、金額　　　　　根據信用狀第　　　　　號
We send you herewith for negotiation our draft No. _____ for _____ drawn under L/C

開狀銀行：
issued by _____

(予 場)

及下列各項件即請查照並希准予辦理押滙為荷：
accompanied by the following documents:

Expiration Latest Date- shipment 　　　　　presentation Documents required by L/C are duly marked 〔 ✓ 〕 滙票 〔 〕Draft〔 〕 商 業 發 票 〔 〕Commercial invoice〔 〕 提　　單 〔 〕Bill of Lading〔 〕 海關發票 領事發票 〔 〕Customs/Consular invoice〔 〕 產 地 證 明 書 〔 〕Cert. of Origin〔 〕 保　險　單 〔 〕Insurance Cert〔 〕 包 裝 重 量 單 〔 〕Packing/Weight List〔 〕 檢　驗　書 〔 〕Inspection Cert.〔 〕 出 口 商 兩 件 〔 〕Shipper's Statement〔 〕 轉 讓 書 〔 〕Letter of transfer〔 〕 船公司證明 〔 〕Carrier's cert.〔 〕 電 報 副 本 〔 〕Copy of cable〔 〕 其 他 單 據 〔 〕Other documents: 郵 政 收 據 〔 〕postal receipt〔 〕 保險公司證明 〔 〕Insurance Co's cert.〔 〕 本筆押滙款項，請依下列方式處理 〔　〕請即折換新台幣墊付貨款存入 　　　活存戶 _____ 　　　支　存 _____ 　　　開支票 _____ 〔　〕還外銷貸款 　　　L C No. 　　　(or Order No.) 〔　〕請扣除費用後即存入外存款 　　　帳戶 〔　〕俟實際收付貨款全額後存入外 　　　滙存款帳戶	**CHECK MEMO** 核　　准 本業各項滙票經核與信用狀條款相符擬請准予承做 副　襄　科　副　經 理　理　長　付　辦 　　　　　長 本業各項單滙經核與L C規定不符如下，擬請准憑L/I承做： 備 註：

85.11.5,000 F

東、上項款請依照外滙管理辦法規定結付。
本公司負責保証決不使貴局墊付上項票據而共遞交付拒方，上項單據如發生違章拒付、
罰 鍰，不 論 局 追 索 全 額 全 數 或 一 部，本 公 司 即 接 照 原 幣 加 息，隨 即 歸 墊 如 有 償 還
額 負 擔 一 切 因 此 而 支 出 之 費 用。
本公司同意因單據上之大地眼疵或因單據正由貴局審核中找不能及時墊付上項票據，
而使本公司同滙率之變動而蒙交損支時，由本公司有一切責任概與貴局無關。

　　　For the proceeds, please have it by settling exchange as stipulated by the regulations governing foreign exchange transaction.
　　　In consideration of your negotiating the above mentioned documentary draft, we undertake to hold your harmless and indem-
nified against any discrepancy which may cause non-payment and/or non-acceptance of the said draft, and we shall refund you
in original currency the whole and/or part of the draft amount with interest and/or expenses that may be accured and/or incurred
in connection with the aboved upon receipt of your notice to that effect.
　　　We further make it known that we agree to stand any loss which may occur through fluctuation of the exchange rates during
the time you are checking the documents before negotiation or consequent on delays in negotiation upon your discovery of some
shortfalls or discrepancies in the documents, and we undertake that you will not be held responsible for any such losses.

地　　址
Address:

電話
Tel:

Yours faithfully

負責人簽章　　Authorized Signature

〔例 9－19〕

①
Exchange for USD28,800.00

② Taipei, Taiwan OCT 31 1996

③ At XXXXXXXXXXXXX sight Pay this *FIRST* of Exchange (Second being unpaid) to the Order of ④

CENTRAL TRUST OF CHINA, FOREIGN DEPARTMENT

⑤ The sum of US DOLLARS TWENTY EIGHT THOUSAND EIGHT HUNDRED ONLY

⑥
Value received in reimbursement of drawing under Letter of Credit No. 31-0274

Dated 961007 Issued by THE BANK OF KYOTO, LTD.

FOREIGN DEPT.

⑦
To BOKFJPJZ

⑧ F H F

1

〔例 9－20〕

台灣地區廠商辦理大陸出口台灣押匯申報表

本次出口有關資料：

(1)目的地國別：

(2)產品之貨品號列及金額

產 品 之 標 準 分 類 號 列 及 檢 查 號 碼（共 十 一 碼）	金　　　　　　　　　　　額

※請依照中華民國商品標準分類號列填列，若有兩種以上，
　請分別填列，若空格不敷使用，請另以紙浮貼。

　　　此　　　致

中 央 銀 行 外 匯 局

（下列資料由指定銀行填寫）

銀行編號：＿＿＿＿＿＿＿＿＿＿＿

押匯託收金額：＿＿＿＿＿＿＿＿＿

指定銀行簽章及日期：＿＿＿＿＿＿

中央信託局

申報人簽章：＿＿＿＿＿＿＿＿

營利事業
統一編號：＿＿＿＿＿＿＿＿

地　　址：＿＿＿＿＿＿＿＿

電　　話：＿＿＿＿＿＿＿＿

傳　　真：＿＿＿＿＿＿＿＿

第一聯：送外匯局（作為出口結匯證實書附件）

〔例9-21〕

損害賠償約定書
LETTER OF INDEMNITY

中央信託局外匯業務處公鑑：
CENTRAL TRUST OF CHINA
FOREIGN DEPARTMENT
TAIPEI, TAIWAN 100

日期
Date:

Dear Sirs,

　　茲為請求 貴處墊付本公司所開匯票第　　　　　號
In consideration of your negotiating our Draft No.

付款人
Drawn on

金額
for _____ under Letter of Credit No.

係 屬 信 用 狀 號 碼

開證銀行名稱
Issued by

鑒於原條款規定
Which stipulates

與 所 提 有 關 文 件 內 容 不 符
Whereas the relative documents indicate

茲 本 公 司 保 證 設 若　　　貴 處 以 墊 付 上 項 與 信 用 狀 條 款 不 符 之 匯 票
We hereby undertake to indemnify you for whatever loss and/or damage that

致 遭 受 損 害 時 當 由 本 公 司 負 責 全 數 償 還
you may sustain due to the above-mentioned discrepancy (ies)

Faithfully yours,

立 約 人 簽 章

85.11.2,000 F

地　址：
Address:

練習問題 •────────────────────────

一、何謂信用狀交易，試述其交易過程。

二、外銷貨品在未裝運之前，有那些作業待辦？

三、貨櫃運輸的方式有幾種？

四、試述裝運通知應有的內容。

五、試述商業發票的主要內容。

六、試述裝箱單的主要內容。

七、試述海運提單的主要內容。

八、試述保險單據的主要內容。

九、〔例 8－2〕的信用狀，其受益人在作業上應注意事項爲何？

十、〔例 8－3〕的信用狀，其受益人辦理押匯時，應提示何種單據？

第十章 Trade English

提貨與服務

一、進口贖單

㈠信用狀項下贖單

出口商所提示的匯票及有關單據，經押匯銀行受理押匯後郵寄開狀銀行。開狀銀行接獲此項單據，經核與原開發的信用狀條款相符者，即繕製到單通知書〔例 10 −1〕通知進口商前來贖單。若到單有瑕疵，則在通知書上列舉瑕疵，而進口商得於合理時間內表示接受與否。

1.全額結匯案贖單手續

進口商於開發信用狀時，已將信用狀金額全數結匯者，憑通知書加蓋原印鑑領單。

2.部份結匯案贖單手續

(1)進口押匯：開發信用狀時與開狀銀行訂定進口押匯契約者，到單時除加蓋原印鑑於通知書上外，應將開狀銀行到單通知書上的到單金額，即匯票金額〔例 10 −2①〕，減去可抵用原繳保證金〔例 10 −2②〕後的墊款金額〔例 10 −2③〕，以還款當日之賣出匯率折合新台幣償還本金，並自國外押匯日起算至還款日為止，按約定利率計算墊款利息，償付開狀銀行後領取單據。分批到單時，保證金原則上成比例抵用。

(2)購料貸款：購料貸款項下開發信用狀者，則可提領單據並自國外付款日起算 180 天內於約定到期日還款，償還本金並繳付利息。

(3)遠期信用狀：遠期信用狀項下的單據，進口商於接獲開狀銀行的通知書後，若係買方遠期信用狀(Buyer's Usance)，即貼現利息由進口商負擔者，承兌匯票後領單，並於到期日償還票款及利息。若係賣方遠期信用狀(Seller's Usance)，即出口商負擔貼現利息者，憑承兌匯票領單，而到期日償還票款即可。

〔例10-1〕

中央信託局外匯業務處
CENTRAL TRUST OF CHINA
FOREIGN DEPARTMENT
49, WU-CHANG ST. SEC. 1
TAIPEI, R. O. C.

到 單 通 知 書

☐ COLLECTION
☐ CLAIM

Taipei,

承辦人員
代 號

TO		Drawn Under L/C No.	Amount of Draft

Their Ref. No.

Negotiated by
Reimbursed on   Date :

Drawer	Drawers No.	Date of Draft	Commission	Postage	

Delivery Date	Invoice	Pkg List	B/L AWB	Insp. Cert.	Mailing Cert.	Tlx Copy	Insurance Cert.	Origin Cert.	Supplier Cert.	

Discrepancies:

一 為保障　貴方權益，務請依照信用狀統一慣例規
　定，於合理期限內審慎是否接受上述瑕疵。
二 檢附本案匯票乙套，敬請查收，並煩速簽回本本
　通知聯。
※ 收到匯票請惠予加蓋章，如有任何問題，務請於
　一日內通知本處處理。

☐ 查本案業於　　年　　月　　日應　貴方之申請辦
　理擔保提貨 ／ 提單背書手續，依照申請書約定
　貴方應無條件同意接受本單據並支付貨款，其有關
　單據已寄達本處，茲隨函附上，敬請查收。

Yours faithfully
CENTRAL TRUST OF CHINA
FOREIGN DEPARTMENT

上列單據業經照收無誤；瑕疵全部接受，此據。
日　期　　　　　　顧客簽章

〔例10-2〕

中央信託局 外匯業務處
CENTRAL TRUST OF CHINA
FOREIGN DEPARTMENT
49, WU-CHANG ST. SEC. 1.
Taipei, R. O. C.

到 單 通 知 書
☐ COLLECTION
☐ CLAIM

承辦人員
代 號

Taipei,

TO	Drawn under L/C No.	
	憑 信 用 狀 號 數	
	Draft or Receipt Amount	
	匯 票 或 收 據 金 額	
	% Draft or Receipt Amount Advanced by CTC	
	本 局 墊 款 金 額	
	Margin used	
	保 證 金 抵 用 金 額	

Negotiated by
Reimbursed on ....

Their Ref. No.
Date:

Payment Commission	Advising Commission	Confirmation Commission		Postage	Total Amount

Delivery Date	Invoice	Pkg List	B/L AWB	Insp. Cert.	Mailing Cert.	Tlx Copy	Insurance Cert.	Origin Cert.	Supplier Cert.

Discrepancies:

二 為保障 貴方權益，務請依照信用狀統一慣例規
  定，於合理期限內答覆是否接受上述取瑕。
三 檢附本裝運證乙各，敬請查收，並傳速簽回正本
  通知聯。
※ 擔保提貨、提單背書以外之案件，收到單據後請
  詳加審查，如有任何問題，務請於一日內通知本
  處處理。

☐ 查本案業於　年　月　日應 貴方之申請辦
理擔保提貨 ／ 提單背書手續，依照申請書約定
貴方應無條件同意接受本單據並支付貨款，其有關
單據已寄達本處，茲隨函附上，敬請查收。

上列單據業經照收無誤；瑕疵全部接受，此據。
日 期　　　　　　　　顧客簽章

Yours faithfully
CENTRAL TRUST OF CHINA
FOREIGN DEPARTMENT

(二)進口託收項下贖單

以進口託收（Inward Collection）方式向國外出口商採購貨品時，進口商於接獲受託銀行的到單託收通知後，依交單方式之不同而分別辦理贖單手續：

1.付款交單（D／P＝Documents Against Payment）：

憑受託銀行的到單託收通知書〔例 10－3〕持往受託銀行結匯領單。貨款則經由受託銀行匯撥至國外託收銀行指定的帳戶。

2.承兌交單（D／A＝Documents Against Acceptance）：

憑受託銀行的到單託收通知書持往受託銀行承兌匯票後領單。匯票的承兌方式通常是由承兌人在匯票正面簽上「承兌」字樣，加註承兌日期並由承兌人簽名。貨款則於到期日再辦理結匯，並經由受託銀行匯撥至國外託收銀行指定的帳戶。

(三)擔保提貨與副提單背書

裝載進口貨物的船舶雖然已到埠，但是信用狀項下有關的單據尚未寄達開狀銀行時，進口商為了早日提貨，以便及早銷售或避免負擔因滯報通關而產生的倉租，得向原開狀銀行申請擔保提貨。

1.向船公司索取並填妥「擔保提貨書」一式兩份〔例 10－4〕。

2.填寫擔保提貨申請書〔例 10－5〕並附商業發票副本、提單副本（Non－Negotiable Copy）及填妥的擔保提貨書。

3.按到貨金額結匯並預繳墊款利息。到貨金額以發票副本為準，並且可將保證金比例抵用後之餘額結匯；墊款利息通常以七天計算，多退少補。

4.領取開狀銀行背書的「擔保提貨書」，憑此向船公司換領提貨單（D／O＝Delivery Order）以便報關提貨。

〔例 10－3〕

1

CENTRAL TRUST OF CHINA
FOREIGN DEPARTMENT
49 WU CHANG STREET SEC. 1
TAIPEI, TAIWAN, R.O.C.

CABLE ADDRESS: "TRUSTEX"
TELEX NR: 21154,23279
S.W.I.F.T.ADDRESS: CTOC TW TP

In reply please quote
Our Ref.

Item received for coll. from

Taipei,

TO

Their Reference No.

Dear Sirs,
Please note that we have received for collection the item mentioned below drawn on you.

Drawer	Drawer's No.	Date of Draft	Due Date/Tenor	Amount

Accompanied by the following documents:

B/L	Invoice	Pack. List	Cert. Orig.	Ins. Cert.	Insp. Cert.

Shipped per S. S.
Covering shipment of
Documents against
Our handling charges   , postages/cable charges
remitting Bank's charges   are for drawee's account.

For CENTRAL TRUST OF CHINA
FOREIGN DEPARTMENT

第十章　提貨與服務

· 211

LETTER OF GUARANTEE FOR PRODUCTION OF BILLS OF LADING

Name of
Consignees: ....

Address: ....

....

Date: ....

TO:WAN HAI STEAMSHIP CO., INC.

TAIPEI.

Dear Sir,

S. S.

M. V .... of **W.H. CONTAINER EXPRESS**

Arrived at .... on .... 19 .... from ....

    In consideration of your releasing for delivery to us or to our order the under-mentioned goods, of which we claim to be the rightful owners, without production of the relevant Bill(s) of Lading (not as yet in our possession).

    We hereby undertake and agree to indemnity you fully against all consequences and/or liabilities of any kind whatsoever directly or indirectly arising from or relating to the said delivery and immediately on demand against all payments made by you in respect of such consequences and/or liabilities, including costs as between solicitor and client and all or any sums demanded by you for defence of any proceedings brought against you by reason of the delivery aforesaid.

    And we further undertake and agree upon demand to pay any freight and/or General Average and or charges due on the goods aforesaid (it being expressly agreed and understood that all liens shall subsist and unaffected by the terms hereof.).

    And we further undertake and agree that immediately the Bill(s) of Lading is/are received by us we will deliver the same to you duly endorsed.

    And we further undertake and agree that no statement relating to the contents, quality, quantity, weight, numbers, marks and/or value of the packages inserted herein shall limit in any way our liability or that of the bankers hereunder.

B/L NO.	MARKS	NUMBER OF PACKAGES	DESCRIPTION OFPACKAGES	CONTENTS	SHIPPED BY

....

Consignees.

    In consideration of your having accepted the above Letter of Guarantee at our request we hereby guarantee the duty performance thereof by the above consignees and agree that no time or other indulgence granted by you to the consignees shall discharge us from our liability hereunder.

Dated .... 19 ......

....

Banker's Signature.

INDEMNITIES WITH LIMITED GUARANTEES OR BEARING ANY
QUALIFYING REMARKS WHATSOEVER CANNOT BE ACCEPTED

0007-A-007
英-09

擔 保 提 貨
副 提 單 背 書　申請書

中央信託局外匯業務處台鑒

逕啓者：

　　茲附奉擔保提貨書／副提單，請　貴處惠予簽署／背書以便向 _____

(船公司)請求提取下列貨品，該貨品係由 _____ (出口港)運抵

(進口港)裝載於 _____ (船名)

信用狀號碼	提單號碼	嘜　頭	貨　品	數　量	金　額

並願遵守下列之約定：

一、貴處因簽署／背書上項擔保提貨書／副提單，致引起之一切後果均由本申請人負責，決不使　貴處因此而蒙受任何損失。

二、同意一俟有關單據寄達　貴處，不論有無瑕疵，本申請人無條件同意接受及授權貴處付款，並請將之以掛號郵寄交申請人，中途如有遺失，概由申請人負責，與貴處無涉。

三、同意擔保提貨後，提單寄達時，即將上項擔保提貨書換回送還　貴處註銷或委由貴處代勞將該項提單逕交船公司換回上項擔保提貨書以便解除　貴處之保證責任。

四、倘若於申請擔保提貨／副提單背書之同時，正本單據已寄達　貴處時，請　貴處同意本申請人以此申請書代替簽回之「到單通知書」領回有關單據。至於該進口單據縱有瑕疵，本申請人亦願意接受不予追究。

申請人

簽署日期　　　　　　　餘額

_____

（請蓋原留印鑑）

簽章人　　　　　　　核對人

5.俟信用狀項下的單據寄達後，由銀行抽出提單正本，向船公司換回擔保提貨書。

信用狀項下的貨物來自航程較近的國家或地區者，爲配合及時提貨，常在信用狀要求出口商將副份提單(Duplicate B/L)逕寄進口商。進口商接到副提單時應向開狀銀行申請副提單背書。

1.填寫副提單背書申請書(同例10－5)並檢附副提單及商業發票。

2.結匯到貨款並預繳墊款利息，結匯款及利息的計算與擔保提貨手續同。

3.領取開狀銀行背書的副提單，憑以向船公司換領提貨單(俗稱小提單)辦理提貨手續。

二、報關與提貨

㈠進口報關

1.報關期限

進口商自開狀銀行贖單及接獲輪船公司到貨通知後，應儘速憑提單向輪船公司換領提貨單並向海關申報進口手續。報關期限則自運輸工具進口日起算十五日內。進口日期的認定方法如下：

(1)海運進口：係自輪船抵達港口後，向海關遞送進口艙單之日起算。

(2)空運進口：係自飛機抵達機場後，海關關員登機查驗並收取進口艙單之日起算。

(3)郵包進口：係自郵局寄發包裹通知單之日起算。

進口商若未依限報關，則自第十六日起，將加征滯報費，滯報費征滿三十日而仍未報關，換言之，自進口日起四十六日後，海關得將該進口貨品變賣，所得價款於扣繳稅捐及必要費用後，如有餘款，由海關代爲保管，進口商在五年以內，檢具提貨單及其他有關證件，向海關申請發還，逾期繳歸國庫。

2.報關文件

進口貨品由納稅義務人或受委託之報關行向海關遞送進口報單申報。進口報單由報關人據實填報正本一份及副本數份(視需要加繕)並檢附下列文件：

(1)提貨單(D／O＝Delivery Order)或空運貨單(AWB＝Air Waybill)影本一份。

(2)商業發票二份。

(3)裝箱單或重量尺碼單一份。

(4)輸入許可證第三聯(正本)，免簽證項目者免繳。

(5)委託書一份，用於確定報關行之委任關係。

(6)貨價申報書二份：進口免徵關稅等情形者免繳。

(7)產地證明書：國貿局規定應提供者，或海關認為有必要或查驗認定生產地不易時得請納稅義務人提供產地證明。

(8)其他：如自美國進口蘋果、葡萄、橘子、李子等鮮果，應檢附美國農業部檢驗證明。進口轎車則應加附「進口汽車應行申報事項明細表」等。

3.通關程序

我國海關隸屬於財政部，其最高之行政機關為關稅總局(The Directorate General of Customs)掌理全國海關行政。關稅總局下轄基隆、台北、台中及高雄等四關稅局。在業務性質方面，台北關稅局以空運貨物、出入境旅客、郵包等業務為主，其他各局均以海運貨物為主。通關程序大致為收單、建檔、查驗、分類估價、繳稅及放行。

(1)收單：投遞進口報單時必須依海關轄區劃分，向貨物存放地主管單位辦理。有電腦連線者先用「電腦傳輸」申報內容給海關，再補遞送書面報單等。未連線者持書面報單等遞交海關後，由海關據以輸入申報內容，其後續之流程則與連線者相同。

(2)建檔：未連線者將書面報單遞交海關，由海關鍵入。如申報資料不正確，電腦無法入檔，可立即退件更正。

(3)查驗：應先由納稅義務人申請查驗，海關始予派人辦理。申請期限自報關日起十日內。並在海關規定之時間及地點起卸查驗。信譽良好之生產事業整櫃貨物得申請在船邊抽驗放行。驗貨方式分①簡易查驗；②一般查驗及；③詳細查驗。查驗時納稅義務人應①負責搬移、拆包或開箱、恢復原狀等事項及費用；②提供參考資料，如型錄、說明書、藍圖、圖樣、成份表等；③簽認查驗結果及取樣事實。

(4)分類估價：在這一步驟，海關辦理稅則分類、核定完稅價格、核銷輸入許可證、未申報與虛報之處理及核計稅捐等項。所謂完稅價格(DPV ＝Duty－Paying Value or Duty－Paid Value)乃關稅、貨物稅、商港建設費、公賣利益、推廣貿易服務費及營業稅核計之基礎。完稅價格最後之核定，係由關稅總局驗估處為之；惟進口小額樣品則授權各進口地海關自行核估。實際進口之貨物經海關查核結果，與原許可之貨物相符者，海關即在許可證上紀錄進口數量及其金額。經准予紀錄而未加註特殊說明者，即完成銷證工作。其進口之貨品即為合法之進口貨品。

(5)徵稅：計有關稅、商港建設費、貨物稅、營業稅、推廣貿易服務費、公賣利益、滯報費、滯納金及規費等。關稅應自海關填發稅款繳納證之日起十四日內繳納。應繳納關稅之進口貨物，於繳納關稅後予以放行。惟經海關核准已提供擔保者，得先予放行。

(6)放行：經過海關核發稅款繳納證，並查核其他機關委託代查之文件，如輸入動物檢疫證明書(Veterinary Certificate)、輸入植物檢疫證明書(Phytosanitary Certificate)、輸入檢驗合格證書(Certificate of Import Inspection)或書刊驗放通知單等之後，由電腦自動列印「進口貨物電腦放行通知」，或傳輸訊息給報關行及貨棧。納稅義務人則於完稅後放行之前向貨物稅證照股辦理核發貨物稅完稅照。

㈡提貨

進口貨物憑海關蓋印放行並經駐棧關員簽章之提貨單核對船名、船次或班機航次與貨物之標記號碼或航空標籤號碼及件數無訛後始得提貨出棧。此際所謂貨棧(Warehouse；Godown)，依海關管理進出口貨棧辦法，乃指經海關核准登記，專供存儲未完成海關放行手續之進出口或轉口貨物之場所。

三、貿易糾紛(Trade Claim)

一筆對外貿易，自進行、成交及至履約的各階段中，任何一方的疏忽或錯誤，可能導致糾紛。此項糾紛由買賣雙方洽商解決的稱為貿易糾紛。遭受損失的一方，為了主張權利或取得相當的代價，先是抱怨(Complaint)，繼而有所爭執(Dispute)甚至要求賠償(Claim)。無論如何解決，貿易糾紛宜採取友好的方式(Amicable Settlement)或交付仲裁(Arbitration)，除非萬不得已，否則不輕言對簿公堂，以免傷感情。

㈠對逾期交貨要求賠償(Claim on delay in shipment)

實踐貿易公司於六月間外銷美國的游泳裝一批，因疏忽交貨期而遭到對方來信抗議，並提出減價的要求〔例 10－6〕。

對外貿易事關所交貨品的品名、數量及規格等，如在訂單上已有詳細記載者，買賣雙方在往來的書信中，常以訂單號碼代之。本函主題與所裝載貨物有關，故而標出主旨引起對方的注意。

主旨：關於本公司第十五號訂單項下泳裝事宜。

第一段：通知船到日期，並提醒對於未按約定期限交貨的後果。

1. the subject goods：上述貨品，即指主旨所提的貨品。類似的表達有 the captioned goods。

2. for some time past：前曾，不久以前。

〔例 10-6〕

Gentlemen:

Re Our Order No. 15, Bathing Suits

The subject goods reached here on July 10, 1998 by M. V. " California ", but we much regret to have to inform you that we cannot accept them.

For some time past we have been pressing you for the shipment of these goods, and in our last letter dated May 31, 1998, we informed you that unless this order was already on the way, the parcel would arrive too late for the season and so be of no use to us.

Nevertheless, you have sent us the goods, and now the only thing we can do is either to take them on consignment or to accept them for next season at a reasonable allowance.

We ask for your immediate decision in the matter, and in the meantime we hold the goods subject to your reply.

Yours sincerely,

---

3. press for：催促，屢次要求，we have been pressing you for 本公司一直在催促請貴公司……。

4. this order：這批所訂貨品。

5. too late for the season：不合時宜，無法應景。

文意：裝載上述貨品的加州輪業於七月十日抵達此地。惟本公司歉難接受該項貨品，蓋前些時日，本公司曾經一再催促貴公司裝運，並以五月三十一日最後一次信轉知貴公司，除非本批貨業已裝出，否則所到貨品，不合時宜，亦毫無用處。

第二段：但是貨品現已運到，苛責無濟於事，唯有提出可行的解決辦法才是上策。

1. Nevertheless：雖然如此 (not withstanding)。

2. take them：指 take delivery of them 提貨。

3. on consignment：寄售方式。

4. allowance：減價，折價。賣方爲所交貨物的損壞、短少、不符品質、遲延交貨等理由而作的減價。成交時以減價優待則爲折扣discount。

　　文意：貴公司不理上述要求而裝出貨品，本公司惟有以寄售方式提貨或作適度降價，方願領貨供下季銷售之用。

第三段：以等候出口商的回答作爲結尾。

1. in the meantime：當前，目前。

　　文意：請貴公司即時作決定，在未獲答覆之前，本公司暫時保留貨品。

(二)答覆索賠(Response to claim)

　　出口商於接到進口商的索賠信之後，應立即調查引起糾紛的原因，若己方的疏忽或錯誤，宜心平氣和接納對方的指責，並作合理的補償。倘糾紛的原因不在我方，則具體地答覆。

〔例10-7〕

Gentlemen:

With reference to your letter dated September 5, 1998 concerning two cases of defective goods shipped under Order No. 22, we would like to state that after our careful investigation, we could not find any error on our part since every effort was fulfilled prior to shipment as shown in the enclosed Inspection Certificate.

Moreover, the goods were on board the ship in perfect condition as clearly stated in the clean B/L. Therefore, we suggest you lodge a claim with the insurance company. We will assist you wherever possible to process the claim.

We are very sorry for the inconvenience you have suffered and assure you that we will make every attempt to prevent such a recurrence in the future.

Yours faithfully,

實踐貿易公司在九月間接獲進口商的索賠函，經過追蹤調查的結果，顯示非我方的責任乃婉覆如上〔例 10－7〕。

　　第一段：覆對方來信，陳述調查的結果。

　　1. defective goods：有缺點的貨，有毛病之貨。

　　2. investigation：調查。出口商接獲進口商的抱怨或索賠信件之後，須自行做內部的追蹤檢查。倘若因此能發現不妥或缺點，當可藉機改善，此際應心平氣和接受對方的指責，並提出妥善的處理辦法或承擔賠償。惟經過細心的查對後，責任非屬我方時，亦應提出具體的理由婉覆對方。

　　3. on our part：屬於我方，本公司方面。

　　4. every effort：一切努力，盡力。

　　文意：謹覆貴公司九月五日大函，有關第 22 號訂單的貨品之中兩箱有問題乙節，經細心查究結果，本公司已盡力完成交貨並無疏忽之處，隨附檢驗證明書為證。

　　第二段：到貨之缺點其責任不在出口商，則貨物的毀損可向承運人 (Carrier)、承攬運送人 (Forwarding agent) 或保險公司 (Insurance Company) 請求損害賠償的索賠。實踐貿易公司於是建議進口商向保險公司索賠。

　　1. lodge a claim with. . . . ：向…索賠，lodge 提訴狀、控訴。

　　2. process the claim：進行索賠，訴訟索賠。

　　文意：再者，本公司所交貨品，裝船時一切尚屬完整，有清潔提單為證，因此謹建議貴公司向保險公司索賠，本公司當盡力協助貴公司進行索賠。

　　第三段：雖然本案錯不在我，唯以安慰話作為結尾。

　　1. the inconvenience you have suffered：貴公司所遭遇的困擾，貴公司的麻煩。

　　2. recurrence：再發生，復現。

　　文意：對貴公司遭遇的麻煩，本公司甚以為憾，但願能盡力避免再度發生類似的情事為禱。

(三)承擔賠償(Adjusting claim)

　　實踐貿易公司向美國進口零件，因規格不符去函抗議並要求折價三成。對方發覺誤裝貨品，於是來函接受賠償要求〔例10-8〕。

〔例10-8〕

Gentlemen:

Thank you for calling our attention in your letter dated October 2, 1998 to the inferior goods shipped under your Order No. 107.

Upon tracing, we find that, through the error of our shipping department, we shipped the goods of Pattern No. AC-12 instead of those of Pattern No. AC-11 on which your order was placed. We sincerely regret we have much troubled and inconvenienced you through our oversight.

In order to adjust the matter, we ask you as you propose, to accept the goods at a reduction of 30%, thought it is a great loss to us.

We appreciate your having given us the opportunity to correct our fault in this business and will do our best not to cause you similar inconvenience again.

Yours sincerely,

第一段：感謝對方來信提醒錯誤的發生。

1. call one's attention to：促請注意，提醒。

　　本文將形容詞片語 in your letter dated October 2, 1997 插入 attention 與 to 之間。

2. inferior goods：不良貨品，劣質品。inferior 為 superior 的反義字。

文意：謝謝十月二日大函提醒第一〇七號訂單的貨品品質不符。

第二段：告知誤裝情形並致歉意。

1. trace：追蹤、追查。

2. trouble：使麻煩，使困擾。

3. inconvenience：使不便，使不愉快。

4. through oversight：由於失察，因疏忽。

文意：經查本公司的交貨部門，將貴公司所訂 AC－11 型的貨誤裝 AC－12 型。由於此項失察，致使貴公司諸多困擾不便之處，本公司甚覺歉意。

第三段：接受三成折價的要求。

1. adjust：調整，整理。

2. propose：提議。

3. at a reduction of 30%：以三成的折價。

文意：為處理本案，擬照原議折讓三成，尚請體諒。

第四段：擔保不再有類似錯誤作為結尾。

1. fault：錯誤，過失。

2. do our best：盡全力。

3. not to cause：不使之發生。

文意：謹謝承蒙貴公司指教改正，本公司當盡力以免再錯。

練習問題

一、簡述信用狀項下贖單之方式。

二、簡述進口託收項下贖單之方式。

三、進口商接獲貨物到埠通知，但是無正本運送單據時，如何始可先行辦理提貨手續？試簡述其辦理方式。

四、進口商進口貨物應於何時報關？

五、報關時須附送的文件中，與進口贖單相關之文件有幾項？

六、簡述我國海關之組織及其業務性質。

七、簡述進口通關程序。

八、試述 Claim 的種類。

九、由國外進口之貨品品質太差,你如何處理?

十、細讀下列的 Preliminary Notice of Claim 並研究各項問題。

　　1.試指出本函的當事人。

　　2.試述本案的情況。

　　3.爲何繕寫本函?

　　4. c. c. 應以何方爲副本抄送單位?

　　5.有無附件?爲何?

　　6.對於本案將來的發展情況及結果如何,試述你的見解。

Dear Sirs,

<u>Notice of Claim No. IM – 2</u>
African Phosphate Rock
<u>Under B / L No. 1087, Inv No. F – 355</u>

　　The Captioned cargo has been found badly contaminated due apparently to the collapse of the stowed heap of Rocks off T / D into the unclean lower hold, and we have applied to the Far East for immediate survey as agreed upon over the telephone.

　　Meanwhile, we hold you responsible for the contamination damage and reserve the right of filing a claim with you pending receipt of the Survey Report.

　　Please acknowledge receipt of this Notice.

Yours faithfully,

c. c.
Enclosure:

附 Trade English 錄

附錄一、出進口廠商登記管理辦法

中 華 民 國 八 十 二 年 七 月 九 日
經濟部經⑻貿○八六七四二號令發布
中 華 民 國 八 十 三 年 元 月 七 日
經濟部經⑻貿○九三一四三號修正

第 一 條　本辦法依貿易法第九條第二項規定訂定之。

第 二 條　公司、行號其營利事業登記證上載明經營出進口或買賣業務,且其資本額
　　　　　(股份有限公司爲實收資本額)在新台幣五百萬元以上者,得依本辦法申請登
　　　　　記爲出進口廠商。

第 三 條　申請登記出進口廠商者,應檢送左列文件:

　　　　　一、申請書及出進口廠商登記卡。

　　　　　二、營利事業登記證影本。

第 四 條　申請登記之出進口廠商,其英文名稱特取部分不得與現有或解散、歇業、註
　　　　　銷或撤銷登記未滿兩年之出進口廠商英文名稱相同或類似。

第 五 條　出進口廠商依法合併或變更名稱、組織、代表人、資本額或營業處所,應於
　　　　　辦妥營利事業登記變更登記後三十日內檢具有關文件向貿易局辦理變更登
　　　　　記。

　　　　　出進口廠商應於辦妥前項變更登記後,始得繼續經營出進口業務。

第 六 條　出進口廠商爲業務需要且無左列情事者,得設置國外辦事處:

　　　　　一、因仿冒商標、標章、僞標產地、侵害專利權或著作權經法院判決確定
　　　　　　　者。

　　　　　二、受暫停出、進口申請之處分,其原因尙未消失者。

　　　　　前項設置國外辦事處應向貿易局辦理報備,其變更時亦同。

第 七 條　出進口廠商依前條規定辦理設置國外辦事處報備時,其應申報辦事處事項如
　　　　　左:

　　　　　一、名稱、所在地址。

　　　　　二、設置年、月、日。

　　　　　三、負責人、籍貫。

　　　　　四、業務項目。

第 八 條　出進口廠商之出進口實績，以海關通關及第九條所核計之實績為準。

第 九 條　出進口廠商所接信用狀轉讓或轉開予其他出進口廠商出口者，得檢附信用狀及經海關簽署證明之輸出許可證或出口報單與其所開立之統一發票等證明文件之影本，向貿易局申請核計出口實績。其辦理上述信用狀之轉讓實績以一次為限。

代理出進口佣金或辦理三角貿易之收入其已報繳稅款並取得有關證明文件者及海外售魚之貨款經行政院農業委員會或各級縣市政府所屬漁業管理單位出具之證明文件者，得併計實績。

前二項出進口實績之核計，貿易局得委託台灣省進出口商業同業公會聯合會、台北市進出口商業同業公會、高雄市進出口商業同業公會辦理。

第 十 條　出進口廠商前一年(曆年)之出進口實績達一定金額標準者，經濟部得予表揚為績優廠商並列入績優廠商名錄。

第十一條　(刪除)

第十二條　出進口廠商如與國外客戶發生糾紛，經貿易局通知者，應於限期內申復。

第十三條　出進口廠商擅自停業或他遷不明者，貿易局得於其原因消失前，暫緩受理該廠商出進口業務。

第十四條　出進口廠商歇業經查明屬實或經命令解散或撤銷登記者，貿易局應撤銷其登記。

第十五條　出進口廠商經撤銷登記者，自撤銷日起或於撤銷前已受暫停處分者自受暫停處分之日起，二年內不得重新申請登記。

第十六條　本辦法自發布日施行。

附錄二、管理外匯條例

中華民國五十九年十二月二十四日總統令公布
中華民國六十七年十二月二十日總統令修正公布
中華民國七十五年五月十四日總統令修正公布
中華民國七十六年六月二十六日總統令修正公布
中華民國八十四年八月二日總統令修正公布

第 一 條　爲平衡國際收支，穩定金融，實施外匯管理，特制定本條例。

第 二 條　本條例所稱外匯，指外國貨幣、票據及有價證券。

　　　　　前項外國有價證券之種類，由掌理外匯業務機關核定之。

第 三 條　管理外匯之行政主管機關爲財政部，掌理外匯業務機關爲中央銀行。

第 四 條　管理外匯之行政主管機關辦理左列事項：

　　　　　一、政府及公營事業外幣債權、債務之監督與管理；其與外國政府或國際組
　　　　　　　織有條約或協定者，從其條約或協定之規定。

　　　　　二、國庫對外債務之保證、管理及其清償之稽催。

　　　　　三、軍政機關進口外匯、匯出款項與借款之審核及發證。

　　　　　四、與中央銀行或國際貿易主管機關有關外匯事項之聯繫及配合。

　　　　　五、依本條例規定，應處罰鍰之裁決及執行。

　　　　　六、其他有關外匯行政事項。

第 五 條　掌理外匯業務機關辦理左列事項：

　　　　　一、外匯調度及收支計劃之擬訂。

　　　　　二、指定銀行辦理外匯業務，並督導之。

　　　　　三、調節外匯供需，以維持有秩序之外匯市場。

　　　　　四、民間對外匯出、匯入款項之審核。

　　　　　五、民營事業國外借款經指定銀行之保證、管理及清償、稽、催之監督。

　　　　　六、外國貨幣、票據及有價證券之買賣。

　　　　　七、外匯收支之核算、統計、分析及報告。

　　　　　八、其他有關外匯業務事項。

第 六 條　國際貿易主管機關應依前條第一款所稱之外匯調度及其收支計劃，擬訂輸出
　　　　　入計劃。

第六條之一　新臺幣五十萬元以上之等值外匯收支或交易，應依規定申報；其申報辦法由中央銀行定之。

依前項規定申報之事項，有事實足認有不實之虞者，中央銀行得向申報義務人查詢，受查詢者有據實說明之義務。

第 七 條　左列各款外匯，應結售中央銀行或其指定銀行，或存入指定銀行，並得透過該行在外匯市場出售；其辦法由財政部會同中央銀行定之：

一、出口或再出口貨品或基於其他交易行為取得之外匯。

二、航運業、保險業及其他各業人民基於勞務取得之外匯。

三、國外匯入款。

四、在中華民國境內有住、居所之本國人，經政府核准在國外投資之收入。

五、本國企業經政府核准國外投資、融資或技術合作取得之本息、淨利及技術報酬金。

六、其他應存入或結售之外匯。

華僑或外國人投資之事業，具有高級科技，可提升工業水準並促進經濟發展，經專案核准者，得逕以其所得之前項各款外匯抵付第十三條第一款、第二款及第五款至第八款規定所需支付之外匯。惟定期結算之餘額，仍應依前項規定辦理；其辦法由中央銀行定之。

第 八 條　中華民國境內本國人及外國人，除第七條規定應存入或結售之外匯外，得持有外匯，並得存於中央銀行或其指定銀行；其為外國貨幣存款者，仍得提取持有；其存款辦法，由財政部會同中央銀行定之。

第 九 條　出境之本國人及外國人，每人攜帶外幣總值之限額，由財政部以命令定之。

第 十 條　(刪除)

第十一條　旅客或隨交通工具服務之人員，攜帶外幣出入國境者，應報明海關登記；其有關辦法，由財政部會同中央銀行定之。

第十二條　外國票據、有價證券，得攜帶出入國境；其辦法由財政部會同中央銀行定之。

第十三條　左列各款所需支付之外匯，得自第七條規定之存入外匯自行提用或透過指定銀行在外匯市場購入或向中央銀行或其指定銀行結購；其辦法由財政部會同中央銀行定之：

一、核准進口貨品價款及費用。

二、航運業、保險業與其他各業人民，基於交易行為，或勞務所需支付之費用及款項。

　　三、前往國外留學、考察、旅行、就醫、探親、應聘及接洽業務費用。

　　四、服務於中華民國境內中國機構之企業之本國人或外國人，贍養其在國外
　　　　家屬費用。

　　五、外國人及華僑在中國投資之本息及淨利。

　　六、經政府核准國外借款之本息及保證費用。

　　七、外國人及華僑與本國企業技術合作之報酬金。

　　八、經政府核准向國外投資或貸款。

　　九、其他必要費用及款項。

第十四條　不屬於第七條第一項各款規定，應存入或結售中央銀行或其指定銀行之外
　　　　匯，為自備外匯，得由持有人申請為前條第一款至第四款、第六款及第七款
　　　　之用途。

第十五條　左列國外輸入貨品，應向財政部申請核明免結匯報運進口：

　　一、國外援助物資。

　　二、政府以國外貸款購入之貨品。

　　三、學校及教育、研究、訓練機關，接受國外捐贈，供教學或研究用途之貨
　　　　品。

　　四、慈善機關、團體接受國外捐贈供救濟用途之貨品。

　　五、出入國境之旅客，及在交通工具服務之人員，隨身攜帶行李或自用貨
　　　　品。

第十六條　國外輸入餽贈品、商業樣品及非賣品，其價值不超過一定限額者，得由海關
　　　　核准進口；其限額由財政部會同國際貿易主管機關以命令定之。

第十七條　經自行提用、購入及核准結匯之外匯，如其原因消滅或變更，致全部或一部
　　　　之外匯無須支付者，應依照中央銀行規定期限，存入或售還中央銀行或其指
　　　　定銀行。

第十八條　中央銀行應將外匯之買賣、結存、結欠及對外保證責任額，按期彙報財政
　　　　部。

第十九條　（刪除）

第十九條之一　有左列情事之一者，行政院得決定並公告於一定期間內，採取關閉外匯市
　　　　場、停止或限制全部或部分外匯之支付、命令將全部或部分外匯結售或存入
　　　　指定銀行、或為其他必要之處置：

一、國內或國外經濟失調，有危及本國經濟穩定之虞。

二、本國國際收支發生嚴重逆差。

前項情事之處置項目及對象，應由行政院訂定外匯管制辦法。

行政院應於前項決定後十日內，送請立法院追認，如立法院不同意時，該決定應即失效。

第一項所稱一定期間，如遇立法院休會時，以二十日為限。

第十九條之二　故意違反行政院依第十九條之一所為之措施者，處新台幣三百萬元以下罰鍰。

前項規定於立法院對第十九條之一之施行不同意追認時免罰。

第二十條　違反第六條之一規定，故意不為申報或申報不實者，處新臺幣三萬元以上六十萬元以下罰鍰；其受查詢而未於限期內提出說明或為虛偽說明者亦同。

違反第七條規定，不將其外匯結售或存入中央銀行或其指定銀行者，依其不結售或不存入外匯，處以按行為時匯率折算金額二倍以下罰鍰，並由中央銀行追繳其外匯。

第二十一條　違反第十七條之規定者，分別依其不存入或不售還外匯，處以按行為時匯率折算金額以下之罰鍰，並由中央銀行追繳其外匯。

第二十二條　以非法買賣外匯為常業者，處三年以下有期徒刑、拘役或科或併科與營業總額等值以下之罰金；其外匯及價金沒收之。

法人之代表人、法人或自然人之代理人、受僱人或其他從業人員，因執行業務，有前項規定之情事者，除處罰其行為人外，對該法人或自然人亦科以該項之罰金。

第二十三條　依本條例規定應追繳之外匯，其不以外匯歸還者，科以相當於應追繳外匯金額以下之罰鍰。

第二十四條　買賣外匯違反第八條之規定者，其外匯及價金沒入之。

攜帶外幣出境超過依第九條規定所定之限額者，其超過部分沒入之。

攜帶外幣出入國境，不依第十一條規定報明登記者，沒入之；申報不實者，其超過申報部分沒入之。

第二十五條　中央銀行對指定辦理外匯業務之銀行違反本條例之規定，得按其情節輕重，停止其一定期間經營全部或一部外匯之業務。

第二十六條　依本條例所處之罰鍰，如有抗不繳納者，得移送法院強制執行。

第二十六條之一　本條例於國際貿易發生長期順差、外匯存底鉅額累積或國際經濟發生重大變化時，行政院得決定停止第七條、第十三條及第十七條全部或部分條文之適用。

行政院恢復前項全部或部分條文之適用後十日內，應送請立法院追認，如立法院不同意時，該恢復適用之決定，應即失效。

第二十七條　本條例施行細則，由財政部會同中央銀行及國際貿易主管機關擬訂，呈報行政院核定。

第二十八條　本條例自公布日施行。

附錄三、輸入許可證各欄填寫說明

欄位	欄 位 名 稱	填 寫 說 明
①	申請人（進口人）	1.申請人（進口人）應依下列格式刻製章戳加蓋（或繕打） 9公分 <table><tr><td rowspan="3">2公分</td><td colspan="2">公司（中文名稱）</td></tr><tr><td>公司（英文名稱）</td><td>營利事業統一編號</td></tr><tr><td>地　　址</td><td>電　話：</td></tr></table> 2.輸入許可證申請人（進口人）名稱不得申請修改，但經貿易局核准變更登記者不在此限。
②	申請人（進口人）印鑑	1.申請人（進口人）印鑑請蓋於貿易局登記之印鑑。 2.如係公營貿易機構代政府機關及公營事業外購貨品進口者，應蓋受託人印鑑以代替申請人印鑑，並請加註「代辦採購人印鑑」
③	生產國別	1.應填貨物之生產國名或地名（進口大陸物品，應繕明 CHINESE MAINLAND，代碼 CN）。 2.右上方格請填填國家代碼（請依「通關作業及統計代碼」手冊規定代碼填註）。
④	起運口岸	係填貨物最初起運口岸之名稱。
⑤	賣方名址	賣方係指報價之國外廠商，右上方格係填寫國家代碼之用（請依「通關作業及統計代碼」手冊規定代碼填註）。
⑥	發貨人名址	1.發貨人係指賣方或其指定之國外廠商。右上方格係填寫國家代碼之用（請依「通關作業及統計代碼」手冊規定代碼填註）。 2.同一份輸入許可申請書申請之貨品，以同一發貨人為限，發貨人不同，應分別填具輸入許可證申請書。
⑦	檢附文件字號	1.進口貨品依規定應檢附主管機關或有關單位文件或（及）特許執照始可申請者，應填明主管機關同意文件或（及）登記證照字號。 2.進口貨品超過一項以上時，主管機關或登記證照字號不同者，請填註證號所屬項次。
⑧	項次	進口貨品超過一項以上時，不論 C.C.C.號列是否相同，均應於項次欄下冠以 1,2,3…，並分別對齊貨品名稱及 C.C.C.號列。
⑨	貨品名稱、規格、廠牌或廠名等	1.貨品名稱應繕以英文為原則，但申請進口中藥材、應加繕中文本草名。貨品名稱不能表明其性質者，應註明其學名。 2.貨品規格係指長短、大小、等級等。 3.貨品名稱欄如不敷填寫，請以輸入許可證續頁繼續填寫，續頁上端請註明共幾項及第幾頁（除最後一頁，可不繕打共幾頁數），並分別加附於各聯之後。 4.除農林漁牧礦、大宗物料等及其他習慣上無廠名或廠牌者可不必繕打外，其他均應繕打 Maker 或 Brand。
⑩	商品分類號列及檢查號碼	指進出口貨品分類表內中華民國商品標準分類號列 CCC CODE 十位碼及檢查號碼。
⑪	數量及單位	為進口統計需要，申請商品 C.C.C.號列第 1 至 21 章，25 至 27 章之農林漁畜等產製品之進口案件，應以我國推行之公制為單位，凡以磅、件、箱、條等為單位者，應折算為公制單位。其他貨品，則依實際使用之單位填列（請依「通關作業及統計代碼」手冊規定代碼填註）。
⑫	單價	1.條件依報價單所載填列如 FOB、CFR、CIF 等 2.進口貨品超過一項以上者，應填列總金額。
⑬	條件金額	3.進口貨品得以新台幣計價，輸入許可證亦可以新台幣報，惟應註明「結匯時以外幣支付」字樣。 4.幣別代碼請依「通關作業及統計代碼」手冊規定代碼填註。 5.不需填列大寫金額。

各聯用途說明：
第一聯：簽證機構存查　　　　　　　　　　　　第三聯：進口人報關用
第二聯：貿易局統計用

附錄四、輸出許可證各欄填寫說明

欄位	欄 位 名 稱	填 寫 說 明
①	申請人（出口人）	1.申請人（出口人）應依下列格式刻製章戳加蓋在該欄 ｜← 9公分 →｜ 公司（中文名稱） 公司（英文名稱）｜營利事業統一編號 地　址　　電　話： （左側：2公分） 2.輸出許可證出口人名稱不得申請修改，但經貿易局專案核准者不在此限。
②	申請人（出口人）印鑑	申請人（出口人）印鑑請蓋貿易局登記之印鑑。
③	目的地國家	係填貨物到達之目的地國家，免填目的地港口，右上角框請填國家代碼（國家代碼請參考關稅總局統計處編印之國別代號表）。
④	轉口港	運輸方式有轉口港者填此欄並加列代號，無則免填。對於限以間接貿易方式出口之地區，則應確實載明轉口港。
⑤	買主	1.他係填列國外買主公司名稱及國別，可免填地址。 2.右上角框填列國家代碼。
⑥	收貨人	如國外收貨人與買主相同，則收貨人欄免填列。　　（第三聯本5、6欄可免填列）
⑦	檢附文件字號	1.出口貨品依規定應檢附主管機關或有關單位文件者，應填明該文件字號。 2.其他規定須加註事項，亦應於此欄說明。
⑧	項次	出口貨品超過一項以上時，不論C.C.C.號列是否相同，均應於項次欄下冠以1,2,3…並分別對齊貨品名稱及C.C.C.號列。
⑨	貨品名稱、規格等	1.貨品名稱應繕打英文為原則。 2.貨品出口簽審規定須填列製造商者，亦應於此欄載明製造商。
⑩	商品分類號列	商品分類號列係指進出口貨品分類表內中華民國商品標準分類號CCC CODE十位碼及檢查號碼。
⑪	數量及單位	使用之數量單位，應依據現行進出口貨品分類表內該項貨品所載之單位填列，如實際交易之數量單位與該數量單位不同時，則於實際交易數量單位下以括弧加註經換算之數量單位。
⑫	條件及金額	1.條件係指交貨條件如 FOB、C&F、CIF 等。 2.總價係填列出口貨品之單項總價及所有貨品之總價。 3.出口貨品得以新台幣計價，惟國外支付貨款仍應以等值之外幣為之，輸出許可證載明外幣，新台幣部分以括弧加註。

注意：⑧～⑫欄如不夠填寫，請以輸出許可證續頁繼續填寫，續頁上端請註明共幾頁及第幾頁，並分別加附於各聯之後。
各聯用途說明：
第一聯：簽證機構存查用
第二聯：簽證機構送貿易局統計用
第三聯：出口人報關用

附錄五
貿易書信常用基本字彙

一、當事人 The Parties Concerned

1. 進出口商 importers & exporters

     trading firm

     trading company

     殷實(有信用)的 reliable

     有經驗的 experienced

          with a wealth of experiences

     名符其實的 a bussiness houses of good records and reputation

2. 製造廠商 maker(s), manufacturer(s)

3. 買方 buyer(s), purchaser(s)

4. 賣方 seller(s), vendor(s)

5. 貨主 shipper(s)貨品的託運人，通常指出口商

          consignee 卸貨地的受貨人，進口報關時的進口商

6. 供應商 supplier(s)通常指提供貨品的出口商

7. 顧客，客戶 a customer, a buyer; a client, a connection

8. 代理商 agent(s)

     總代理 sole(exclusive)agent(s)

     總經銷 sole distributor(s)

     代理權 agency

     代理佣金 agency commission

     代理契約 agency agreement(contract)

     做代理商 act as the agents for one, as one's agents

     總公司(授權代理的一方)principals

9. 同業 competitor(s)

10. 貿易商，商社 a trading firm

11. 業界 business circle

12. 分公司 branch offices

13. 本公司 we, our company(firm), this company(firm), ourselves

     由本公司 on our side, at our end

14. 貴公司 you, your company(firm), yourselves

由貴公司 on your side, at your end

15. 該公司 they, themselves

　　由該公司 on their side, at their end

二、詢　價 An Inquiry

1. 向……詢價 send(make, give)one an inquiry

　　受理詢價(應對方的詢價)entertain an inquiry

　　詳細的詢價 specific inquiry

2. 有意(感興趣買)an interest, if interested

　　無意(無興趣買)be uninterested

3. 買賣 business

　　交易 a deal

　　有希望的交易 business in possible(feasible)

　　成交、訂約 conclude a contract

　　未成交 business has fallen through

　　可觀的交易 fairly good business, sizable business

　　實績 business performance(records, showing)

4. 貨品 commodities, goods, merchandise

　　產品 a product

　　項目 an article, item

　　經手 handle, deal in

　　台灣製 Taiwan(Taiwanese)make

　　貨物 cargo

5. 種類 kind, sort, nature, variety, type, pattern, style, design, grade

6. 品質 quality

　　品質保證 quality guarantee

　　品質檢查 quality inspection

　　品質不良 the quality is poor, the goods are of poor(inferior)quality

　　檢查 check, examine

7. 規格 standard, specifications

　　清單 specifications

8. 樣本 sample(s), specimen(樣式)

   (布匹等剪下的)樣品 sample cutting(s)

   一系列的商品樣本 a full range of samples

   免費樣品 free sample

   裝船樣品 shipping sample

   相對樣品 counter sample

9. 目錄 catalogue(s), catalog(s)

   小冊子 leaflet(s), booklet(s), pamphlet(s), brochure, literature

   圖解 illustration, sketches

   設計圖 plan, drawing, blue print

10. 價目表 a price list

   標價(價目表上的)listed price(s)

   (價格)昂貴 too high, incompetitive, unworkable

   (價格)公道 low, competitive, workable, reasonable

11. 報價 a quotation

   報價 quote(a price)

12. 備咨人 reference

   備咨銀行 bank reference

   備咨商號 trade reference

   本公司的備咨人是台灣銀行 Our reference is Bank of Taiwan.

   徵信資料 credit information

13. 需要 demand

   實際使用者 ultimate consumer, end user

   大量需求 brisk demand, heavy demand

   國內消費 domestic consumption, home demand

14. 供給 supply

   供給品 supplies

   缺乏供給 supplies are scarce

   由於供應不濟 due to scarcity(shortage)of supplies

   由於供應過多 due to abundant supplies(overstock)

15. 市場 a market(市況)

買方市場 a buyer's market = a market in favor of the buyers

賣方市場 a seller's market

市場調查 market research(survey, inspection)

市場報告 market report(information)

開拓市場 cultivate a (new) market

市場趨勢 a market feeling(trend, inclination), the run of the market

往市場採購 be in the market

往市場銷售 to be marketed, be put on the market

16. 銷售 sale

銷售路線 sales, outlet(s), sales(distribution) channels

銷售展望 sales prospect

看好 a promising (rosy) prospect

無望 poor (little) prospect

17. 風險 a risk

風險大 risky

風險由貴公司承擔 at your own risk (and account)

殆無風險 scarcely with any risk

18. 選擇權 an option(由兩個以上的條件中擇一的權利)

賣方的選擇權 at seller's option = S. O.

買方的選擇權 at buyer's option = B. O.

19. 佣金，酬勞金 Commission

手續費 handling charge

銀行手續費 banking charges

費用 charges, fee

免費 free of charge

電報費 cable charge

領事簽證費 consular (invoice) fee

盈餘，利潤 margin

利潤薄 at slim(meagre) margin (of profit)

利潤 profit, returns

好賺的生意 profitable (lucrative) business

三、費　用 Charges（有關輸出入的各項費用）

代墊費用 disbursement（for account of others）

減縮費用 minimize（economize, trim）the charges

費用 fee, charge, expense(s)

支出 expenditure

1. 裝船費用 shipping charges

2. 船上交貨費用 FOB charges

3. 運輸費用 forwarding charge

　　卡車費用 truckage

4. 駁船費 lighterage

　　駁船 a lighter, a barge, craft

　　曳船 a tug – boat

　　小蒸氣船 a（steam）launch

5. 船上裝卸費 stevedorage

　　裝卸人工 stevedores

6. 倉租 storage（包括進倉費 storing charge 及出倉費 unstoring charge）

7. 卸貨費 landing（discharging, unloading）charge

四、價　格 Prices

以交貨地點每一單位的價格，用約定的貨幣表示，例如 US$5. 50 per dozen CIF Los Angeles

1. 單位 unit

　　單位價格 unit price

　　每噸 per ton

　　重量噸 weight ton

　　　長噸 long ton = 2, 240 lb or 1, 016kg

　　　短噸 short ton = 2, 000 lb or 907. 185kg

　　　公噸 metric ton = 2, 204. 6 lb or 1, 000kg

　　lb 為 libra 之簡寫 = pound

　　體積噸 measurement ton

立方公尺 ton of M³(cubic meter) = 35.3147cft

四十立方英呎 ton of 40 cft(cubic feet) = 1.133M³

運費噸(運費以重量噸或體積噸為單位者)

W/M = by weight or measurement

= Freight ton, shipping ton

2. 基本價格，成本 cost

船上交貨成本 FOB cost

3. 工廠交貨價 Ex Factory(Maker's Godown) Price

4. 火車站交貨價 FOR(Free On Railway) Price

5. 船邊交貨價 FAS(Free Alongside Ship) Price

6. 船上交貨價 FOB(Free On Board) Price

7. 運費在內價 CFR(Cost and Freight) Price

8. 運費保險費在內價 CIF(Cost, Insurance and Freight) Price

9. 運費保險費及佣金在內價 CIFC(Cost, Insurance, Freight and Commission)

Price　佣金百分之三的運費保險費在內價 CIFC₃

10. 進口地船上交貨價 Ex Ship Price

五、各項條件 Terms and conditions

1. 貨品規格 description

項目號碼 item No. (#)

2. 品質 quality

3. 數量 quantity

4. 價格 price; unit price

5. 裝船，交貨 shipment

船期 time of shipment

6. 包裝 packing

出口標準包裝 standard export packing

內包裝 interior packing

外包裝 outer packing(package)

7. 刷嘜 shipping mark

主標誌 main mark

副標誌 submark

側標誌 side mark, caution mark 注意標誌

港口標誌 port mark, destination

產地國名 country of origin

箱號 case Nos(#)

8. 付款條件 payment(terms)

　　憑即期信用狀付款 payment by sight L/C

9. 憑我方確認為有效的條件 subject to our final confirmation

10. 有權先售條件 subject to prior sales, subject to being unsold

11. 檢驗 inspection

　　仔細檢驗 a thorough inspection

12. 代替品 an alternative

六、報　價 Offer

要求報價 invite an offer

報價、發價 offer one an item, make one an offer on an article

取得報價 obtain an offer

未獲報價 an offer is unobtainable

1. 買方報價 a bid, a buying offer

2. 穩固報價 a firm offer = F/O

3. 還價、相對報價 a counter offer

4. 投標 a bid, a tender

　　招標 a tender is invited for an article

　　投標 submit a bid

　　得標 one is awarded(successful) in tender

5. 承諾 accept

　　不接受 hold off, withdraw

　　歉難接受 we regret we would withdraw.

6. 回信，覆文 a reply, response

7. 訂貨 an order, an indent(訂單)

　　訂貨 place an order with one for an article, give one an order

接受訂單 book orders

試訂 a trial order

續訂 a repeat order

8. 有意向 an idea, an indication

可行的價格 possibilities, prices at which business is possible

9. 估價 estimate, proforma invoice（估價單，預估發票）

七、裝　船 Shipment

裝貨、發貨 ship, effect shipment

裝船 load

裝載 lift

裝貨港 port of loading, shipping port

卸貨港 port of discharge, destination（目的地）

提貨 take delivery of the cargo

1. 即期裝船 immediate shipment,（訂約後 1～2 週內裝船）

prompt shipment,（訂約後 2～3 週內裝船）

依 Incoterms 的規定，兩者均以 30 天內為準。

2. 海運運費 ocean freight

運費費率表 freight tariff

以重量或體積計費 payable on weight or measurement（W／M）

基本費率 basic rate

附加費 surcharge

燃料附加費 bunker surchage, BAF（bunker adjustment factor）

港口擁塞附加運費 congestion surcharge

最低運費 minimum freight

運費預付 freight prepaid

運費待收 freight collect

3. 第一船 the first available vessel（ship, steamer）

直航船 a direct steamer

預定進港日 ETA = expected time of arrival

預定開航日 ETD = expected time of departure

貨櫃船 container ship

4. 分批裝船 partial shipments

　　　可分批裝船 partial shipments allowed

　　　不可分批裝船 partial shipments prohibited

5. 轉運 transhipment

6. 裝艙 stowage

　　　裝艙圖 stowage plan

　　　船艙 a hold

　　　艙口 a hatch

　　　艙口蓋 a hatch cover

　　　中甲板 tween deck = T/D

　　　裝載甲板貨 deck cargo

7. 裝載量 loading capacity, shipping quantity

　　　一成上下船方作主 10% more or less at ship's option

8. 艙位 cargo space, space(定期船的艙位)

　　　tonnage, bottom(不定期船的艙位)指載重量

　　　訂艙位 book, arrange, fix, engage freight

　　　艙位不足 shortage(scarcity) of tonnage

　　　包括裝卸費的運費條件 berth terms

9. 租船 charter a ship

　　　傭船 a chartered ship

　　　傭船契約 a charter party = C/P

　　　船主 a shipowner, an owner

　　　運費市場 freight market

　　　FI = Free In 船方不負擔裝貨費用

　　　FO = Free Out 船方不負擔卸貨費用

　　　FIO = Free In and Out 船方不負擔裝卸費用

10. 裝船通知 a shipping advice

　　　裝船指示 shipping instructions

　　　裝貨單 shipping order = S/O

11. 航空貨物 an airfreight

空運 airfreight

八、付　款 Payment

付款條件 payment terms

1. 清償 settlement
2. 信用狀 L/C = Letter of Credit, Credit
   商業信用狀 a commercial L/C
   跟單信用狀 a documentary L/C
   即期信用狀 a sight L/C
   保兌信用狀 a confirmed L/C
   不可撤銷信用狀 an irrevocable L/C
   不可轉讓信用狀 a non – transferable L/C
3. 託收方式 on collection
   付款交單 D/P = Documents against Payment
   承兌交單 D/A = Documents against Acceptance
4. 修改信用狀 amend L/C
   電報修改 cable amend, amend by cable
   展期 extention of L/C
5. 匯票 a bill of exchange, draft(s)
   即期匯票 a sight draft(bill)
   遠期匯票 usance draft(bill), time bill
   　六十天期匯票 60 days after sight bill
   　九十天期匯票 90 d/s bill (90 days after sight bill)
   遠期信用狀 usance L/C
6. 外匯匯率 foreign exchange rate
   買入匯率 buying rate(銀行買入匯率)
   賣出匯率 selling rate(銀行賣出匯率)
   匯率變動 exchange fluctuation
   即期匯率 spot rate
   遠期匯率 forward rate

九、契　約 Contract（合約）

約定，約定書 agreement

1. 訂約 contract, conclucle a contract, make a contract, enter into a contract
   締約 conclusion

2. 更改合約 renew a contract
   更改 renewal（of a contract）

3. 契約條款 stipulations, provisions
   條款 article
   內容 contents

4. 遵照約定，如約 as contracted, as per the contract, as stated in the contract

5. 符合條款 in accordance with the stipulations, according to the terms and con
   ditions stipulated in the contract

6. 代理合約 an agency agreement

7. 銷貨確認書 a confirmation of sale, sales confirmation（在交換正式合約之前，
   或代替合約，由出口商出具確認銷貨內容條件之文書）

8. 履約 fulfilment（execution, performance, carrying out）of a contract
   不履約 non – fulfilment of a contract
   　　　 breach（failure）of a contract
   取消合約 cancellation of a contract
   無效 be nullified, become null and void
   無條件取消 be cancelled unconditionally
   契約餘額 the balance of contract

十、單　據 Documents

1. 商業發票 commercial invoice
   預估發票 proforma invoice
   發票 shipping invoice（裝船後製作的發票）
   領事副簽發票 visaed commercial invoice（經由輸入國駐輸出國之領事加簽的
   發票）

2. 裝箱單、包裝單 packing list

3. 海運/海洋提單 marine/ocean bill of lading（marine/ocean B/L, marine/ocean blading）

   航空運送單據 air transport document

4. 保險單 insurance policy, certificate of insurance, insurance documents

5. 領事發票 consular invoice

6. 檢驗證明書 inspection certificate

7. 產地證明書 certificate of origin

8. 公證報告 survey report

   重量證明書 certificate of weight

   丈量證明書 certificate of measurement

9. 檢疫(衛生)證明書 phytosanitary certificate

10. 特別海關發票 special customs invoice

十一、報　關 Customs Clearance, Customs Entry

1. 通關手續 customs procedures

2. 海關 customhouse, the Customs

   海關關員 a customs officer

3. 關稅，關稅稅率 customs tariff, tariff

4. 進口關稅 import duty

   附加稅 surtax

   課進口稅 impose(levy) duty on one for an article

   免進口稅 be exempted from import duty

   課稅百分之二 a duty is levied at 2%, be subject to a 2% duty

   課稅後每件約值美金五元 it comes up to about US$5. 00 a piece after duty

   從量稅 specific duty

   從價稅 ad valorem duty, adv(ad val.)duty

5. 保稅 in bond

   保稅倉庫 a bonded storage warehouse, a bonded warehouse

   保稅加工倉庫 a bonded manufacturing warehouse

   保稅工場 bonded factory

   保稅貨物 bonded goods

6. 報關行 a customs broker

貨運承攬業者 a forwarding agent, a freight forwarder

十二、索 賠 Claim（抱怨，求償）

1. 抱怨 a complaint, a protest

申訴抱怨 complain , lodge a complaint with one for…，

make protest against one for…

提出索賠 file a claim with one for…

理賠 meet(accept) the claim

索賠通知書 a preliminary notice of claim

保留索賠權利 to reserve the right of filing a claim

品質索賠案 a quality claim

賠償通知 a claim note

2. 爭執、爭論 a dispute

引起爭執 develop into a dispute

提出爭論 bring to a dispute

爭論中 be in dispute

爭執案件 the matter in dispute

3. 損害 damage

損害額 damages

損失 loss(減失)

損失額 losses

運輸途中的損害 damage in transit

破損 breakage

漏損 leakage

污損 contamination(污染)

曲損 bent & dent(彎曲和凹損)

短少 shortage, shortlanded(短卸)

4. 不可抗力 force majeure

5. 檢驗 inspection

出口檢驗 export inspection

鑑定，調查 survey, investigation（索賠階段的調查，檢驗）

鑑定人，公證人 a (sworn) surveyor

申請公證 arrange a surveyor

鑑定中 under survey

十三、海上保險 Marine Insurance

1. 投保 effect insurance

2. 保險金額 insured amount, amount insured

3. 保險費 insurance premium

    保險費率 insurance premium rate

    保險費率表 insurance premium tariff

4. 承保條款 insurance conditions

    承保 insure, cover（投保）

5. 承保範圍 coverage

    保險種類 types of coverage

    全損險 TLO = Total Loss Only

        又稱 FAA = Free from All Average

    平安險 FPA = Free from Particular Average

    水漬險 WA = With Average

    全險 AAR = Against All Risks

        又稱 AR = All Risks

    附加險 Special Clauses

    偷竊及短卸險 TPND = Theft, Pilferage and Non – Delivery

    兵險及罷工暴動險 War & SRCC = War risk & Strikes,

        Riots and Civil Commotion

6. 保險單 insurance policy（保單）

    保險證明書 insurance certificate

    預約保險確認單 open cover

    預約保險單，開口保單 an open policy

7. 保險公司 an insurance company

    保險業者，保險人 underwriter

被保險人 insured

十四、其他各項 Miscellaneous

1. 帳戶 an account, a/c

   通信費帳戶 communication a/c

   電報費 cable charge

   未付款帳戶 outstanding a/c

   應付帳款 account payable

   應收帳款 account receivable

   消帳 strike off the account

   抵帳 offset the account

   收據 a receipt

   銷貨帳戶 account sales

   借方清單 a debit note

   貸方清單 a credit note (進帳單)

   代墊款項 disbursement (for other party's account)

   結轉下期 C. F. = Carried Forward

   上期結轉 B. F. = Brought Forward

2. 寄售貨物 a consignment

3. 存貨 stock

   庫存品 inventory

   庫存過剩 overstock

4. 盤存 inventory (盤存表)

   盤點存貨 stock-taking, make an inventory, take stock

5. 銷貨 turnover

   銷貨量(額)sales

   實績 records

6. 明細 particulars, details

   內容 contents

   細分 breakdown

7. 憑證 voucher(s)(傳票)

附件 enclosure(s)

正本 original

第二份(副本)duplicate

第三份 triplicate

第四份 quadruplicate

收據一式三份 a receipt in triplicate

抄本 a copy

影本 a photostat copy

8. 惠顧 patronage

贊助 support

承蒙惠顧 be patronized

協助，支援 assistant, help, support

承，承蒙 by, through, thanks to, on the strength of

9. 概要 summary

概略 outline

摘要 summarize, sum up（概述）

10. 考慮 consideration（斟酌）

經常（受）照顧 usual courtesy

安排 arrangements

籌備 make arrangements with one for a thing

11. 指示 instructions

遵照指示 as instructed

12. 情況 circumstances

事態 situation（立場）

狀況 condition

13. 瞭解 understanding

在……的條件之下 on the understanding that

得到諒解，達成協議 come to an understanding

理解 comprehension, apprehension

理解 appreciation（感謝）

解釋 interpretation

14. 麻煩 inconvenience(不便)

　　困擾 trouble

　　異議 objection

15. 手段 step(s)(措施)

　　策略 measures

　　採取手段 take steps(measures)

16. 同意 consent(默認)

　　承諾 agreement(同意)

　　承諾 acceptance(接受)

　　認可 approval(承認)

　　取得同意 obtain one's consent

　　說服 persuade one to do

17. 準則 principle, basis, reason

　　作爲原則 as a matter of principle, on principle

　　爲原則 in principle

　　爲依據 on the basis of

　　爲理由 under the reason of

18. 提案 proposal, proposition, suggestion, recommendation(建議)

　　依照提案 as proposed(suggested)

19. 銷售 sell, distribute, market

　　上市 be introduced into the market, be brought to the market

　　暢銷 sell well, enjoy a good sale, be popular, gain popularity

20. 作法 practice

　　方法 ways and means, process

21. 主題 subject

　　要件 matter

　　主旨 purport

　　內容 contents

22. 來函 a letter received from, your letter

　　來電 your cable, a cable received from

23. 拉丁略語

e. g. 例如（exempli gratia）

i. e. 即（id est）

lb 磅（libra）

N. B. 註（nota bene）

per 每

p. a. 每年（per annum）

年息百分之十二點五 interest of 12. 5% p. a.

24. 單位記號

袋 bag Bg

捆包 bale, Bl Bls

木箱 case C／ – C／S

紙箱 carton Ctn Ctns

打 dozen doz

桶 drum Drm Drms

罐 Can Can Cans

個，件，張，本，枚 piece, pc, pcs

捲筒 roll Rl Rls

套 set

碼 yard yd yds

公斤 kilogram kg kilo

公噸 metric ton mt M／T

琵琶桶 barrel（36 加侖）bbl

蒲式耳 bushel（36kg，英斗） bu

加侖 gallon gal.

公升 litre l. lit.

25. 外幣簡寫   SWIFT 用語

美金 US Dollar US$ USD

分 cent c ¢   –

英磅 Pound Sterling Stg £₤   GBP

日圓 Japanese Yen ¥   YEN

澳幣 Australian Dollar A$ AUD

奧地利幣 Austria Shilling AS   ATS

比利時法郎 Belgian Franc BF   BEF

加拿大幣 Canadian Dollar C$ CAD

馬克 Deutsche Mark DM   DEM

法國法郎 French Franc FF   FFR

港幣 Hongkong Dollar HK$ HKD

荷蘭幣 Netherland Guilder DFL   NLG

新加坡幣 Singapore Dollar S$ SGD

南非幣 South African Rand Rand   ZAR

瑞典幣 Swdish Krona SK   SEK

瑞士法郎 Swiss Franc SF   CHF

歐洲通貨單位 European Currency Unit   ECU

紐西蘭幣 New Zealand Dollar NZ$ NZD

泰銖 Thailand Baht B   THB

馬來西亞幣 Malaysia Ringgit M$ MYR

歐元 EURO   EUR

附錄六、認識歐元——外匯業務說明書

中央信託局編

認識歐元

歐元的產生背景

　　歐洲同盟(EU)於 1998 年 5 月決議，由包括德國、法國、荷蘭、比利時、盧森堡、西班牙、葡萄牙、義大利、奧地利、愛爾蘭及芬蘭等 11 個符合條件的會員國成立經濟及貨幣同盟(EMU)，並於 1999 年 1 月 1 日正式啓用歐元(EURO)，作爲 EMU 之單一貨幣。

歐元的轉換過程

一、1999 年 1 月 1 日：

　　(一)歐元之法定地位確立，該 11 個會員國貨幣(National Currency Unit，簡稱 NCU)成爲歐元之記帳貨幣及過渡貨幣，但仍維持法償地位。

　　(二)歐元與各 NCU 之轉換匯率固定。

　　(三)歐元以 1:1 之轉換匯率取代歐洲通貨單位(ECU)。

　　(四)銀行間清算開始採用歐元，新發行公債以歐元計價。

二、1999 年 1 月 1 日至 2001 年 12 月 31 日(過渡期間)：

　　(一)以「不禁止、不強迫」原則逐步轉換 NCU 爲歐元。

　　(二)歐元與 NCU 並行使用，交易雙方可自由議定，選擇歐元或 NCU 作爲計價及交易單位。

三、2002 年 1 月 1 日起：

　　(一)正式發行歐元紙幣及硬幣。

　　(二)7 月 1 日起，歐元成爲唯一法定貨幣，NCU 停止流通。

歐元小檔案

幣別代號	EUR	
最小單位	小數點取至百分位，稱為 cent	
錢幣種類	紙鈔面額	500、200、100、50、20、10、5元
	硬幣面額	2元、1元、50分、20分、10分、5分、2分、1分

歐元的轉換及進位規則

一、歐元與 NCU 之固定轉換匯率以 6 個阿拉伯數字表示(但轉換匯率小於 1 時，小數點前之 0 不計入)。

例：設 1EUR = 1.99899DEM，

1EUR = 0.794019IEP

NCU 與歐元間轉換結果計算至小數點二位，以下四捨五入。

則：100DEM = 100/1.99899

= 50.025262EUR

= 50.03EUR

二、NCU 幣別轉換為另一 NCU 幣別，均須透過歐元，轉換過程歐元至少取小數點三位數，以下四捨五入。

例：設 1EUR = 1.99899DEM，

1EUR = 0.794019IEP

則 100DEM = 100/1.99899

= 50.025262EUR

= 50.025EUR

50.025EUR × 0.794019 = 39.7208IEP

= 39.72IEP

本局外匯業務因應作法

進出口業務

本局進出口業務將依循國際商會 1998 年 4 月 6 日通過之「歐洲單一貨幣(歐元)對國際商會規則(ICC Rules)項下金錢債務交易之影響」決議書辦理，依據該決議書，引進歐元將不影響現行國際商會規則之執行。其主要決議如下：

一、適用於信用狀(包括擔保信用狀)之 UCP500 各種情況及相關規則：

開狀日	付款日	適用幣別
1999/1/1 前，以 NCU 開狀或 1999/1/1 至 2002/1/1 間以 NCU 或歐元開狀	1999/1/1 至 2002/1/1 間	以信用狀幣別付款，開狀行亦可選擇以等值歐元支付，但 1999/1/1 至 2002/1/1 間簽發之單據，提示時得以信用狀幣別或等值歐元或受益人營業處所在國之交叉計算等值該國 NCU 為之。
	2002/1/1（含）以後	須以歐元付款，但 1999/1/1 至 2002/1/1 間簽發之單據，提示時得以信用狀幣別或等值歐元或受益人營業處所在國之交叉計算等值該國 NCU 為之；而於 2002/1/1（含）以後簽發之單據，均須以歐元為之。
2002/1/1（含）以後以歐元開狀	2002/1/1（含）以後	開狀、付款及單據（若係於 2002/1/1（含）以後簽發者）皆須以歐元為之。

NCU：National Currency Unit，指 EMU 加入國之原始貨幣單位。

二、依據前述準則辦理時，同一套提示之單據(包括保險單據)，若以信用狀幣別及/或以歐元及/或以受益人營業處所在國之 NCU 表示者，並不認為單據間彼此抵觸。

三、託收案件須依託收指示書所規定幣別為之。若託收指示書中係規定為 NCU，則自 1999/1/1 起得選擇改用等值歐元支付，而自 2002/1/1 起必須改用等值歐元支付及收受款項。

四、決議書內容亦適用於可轉讓信用狀及保證函。以 NCU 開發之可轉讓信用狀，於過渡期間內轉讓者，轉讓銀行得將其轉換成等值歐元。

匯率及利率掛牌

一、本局目前受理之外匯存款中計有五種為 NCU，包括奧先令(ATS)、比利時法郎(BEF)、馬克(DEM)、法國法郎(FRF)及荷蘭幣(NLG)，於過渡期間仍將維持存款利率及匯率掛牌。

二、歐元與 NCU 之轉換，依 EMU 所定固定轉換匯率計算。

外匯存款業務

一、原 NCU 外匯活期存款：

仍可繼續持有至 2001 年 12 月 31 日或選擇轉換爲歐元活期存款。2002 年 1 月 1 日
前未辦理轉換者，將自動轉換爲歐元活期存款。

二、原 NCU 外匯定期存款：

(一)1999 年 1 月 1 日後到期之存單，仍依原存單上之幣別及利率計息。

(二)1999 年 1 月 1 日後展期之存單，得選擇 NCU 或歐元定期存款。

(三)2002 年 1 月 1 日仍持有 NCU 定期存款者，將自動轉換爲歐元定期存款。

三、原歐洲通貨單位(ECU)之外匯活期、定期存款：

於 1999 年 1 月 1 日以 1:1 自動轉換爲等額歐元存款。

匯出匯款業務

過渡期間可選擇以 NCU 或歐元辦理匯出匯款。

匯入匯款業務

過渡期間以 NCU 匯入之匯款，可選擇以 NCU 或歐元持有，或結售爲新台幣。

買入暨託收國外票據業務

過渡期間 NCU 之國外票據入帳後，可選擇以 NCU 或歐元持有，或結售爲新台
幣。2002 年 1 月 1 日(含)以後入帳者，須以歐元持有，或結售爲新台幣。

中長期貸款業務

NCU 之中長期貸款，合約之權利義務不受歐元實施影響，過渡期間，借款人可選
擇以原 NCU 或歐元支付本金及利息，2002 年 1 月 1 日(含)以後，則以歐元支付。至利
息計算標準，於浮動計息期間未到期前，適用原訂利率水準，新計息期間開始，則改以
歐元利率爲計算基礎。

本說明書僅為概要之說明，於歐元實施後，歐盟、各國政府或主管機關、國際間商會、團體、清算機構以及金融業界等均能繼續制定或發展相關法令、規章、慣例或程序，以補充或修正前述與歐元相關之規範，並影響本局作業程序，特予說明。

國家圖書館出版品預行編目資料

貿易英文：Trade English／李再福著.
--初版.--臺北市：五南，1999 [民88]
面；　公分
I S B N　978-957-11-1775-1（平裝）
1.商業書信　　2.商業－應用文
3.英國語言－應用文
493.6　　　　　　　　88003562

1O25
貿易英文
Trade English

作　　者 － 李再福(82)

發 行 人 － 楊榮川

總 編 輯 － 龐君豪

主　　編 － 張毓芬

責任編輯 － 黃淑真

出 版 者 － 五南圖書出版股份有限公司

地　　址：106台北市大安區和平東路二段339號4樓

電　　話：(02)2705-5066　傳　　真：(02)2706-6100

網　　址：http://www.wunan.com.tw

電子郵件：wunan@wunan.com.tw

劃撥帳號：01068953

戶　　名：五南圖書出版股份有限公司

台中市駐區辦公室/台中市中區中山路6號

電　　話：(04)2223-0891　傳　　真：(04)2223-3549

高雄市駐區辦公室/高雄市新興區中山一路290號

電　　話：(07)2358-702　傳　　真：(07)2350-236

法律顧問　元貞聯合法律事務所　張澤平律師

出版日期　1999年4月初版一刷
　　　　　2011年3月初版五刷

定　　價　新臺幣385元